HONDA ENGINE SWAPS

AARON BONK

S-A DESIGN

CarTech®

Edited By: Travis Thompson

ISBN 978-1-61325-069-3
Item SA93P

Printed in USA

Front Cover: *Jake Johnson's '94 Civic gets fitted with a '97 B18C motor at Hookups Import Tech in Upland, California. The engine produces 195 hp at the wheels, thanks to Integra Type R rods, Civic Type R pistons, some trick head work, Mugen header, a Skunk 2 intake, and a Mugen P28 ECU. (Scott Killeen photo, courtesy of Buckaroo Communications)*

Back Cover, Upper Left: *Some swaps are best suited for a drag car. You'll be hard pressed to get one of these to pass a smog check legally. The guys at Honda Fiend use this CRX strictly for racing purposes.*

Back Cover, Upper Right: *When reinstalling the VSS, be sure that it sits flat before tightening down the bolt. If it doesn't, then the bracket shown on the right of the unit can snap off.*

Back Cover, Lower: *Since this is a USDM engine and this vehicle is already OBD I, smog issues will be taken care of once this vehicle is certified.*

CarTech ®
39966 Grand Avenue
North Branch, MN 55056
Telephone (651) 277-1200 • (800) 551-4754 • Fax: (651) 277-1203
www.cartechbooks.com

TABLE OF CONTENTS

ABOUT THE AUTHOR AND ACKNOWLEDGMENTS

About the Author

Aaron Bonk appeared on my radar in 2000 as a managing partner of Holeshot Racing. The small two-man shop in Anaheim, California reminded me of so many of the speed shops my Dad took me to as a kid. Back in the early seventies, SoCal was busting with little shops like these, each with their own group of followers and regular customers. When cars ran fast, beat other locals, and held together for more than a few weeks, folks got curious about who did the wrenching. If you drove a Honda or Acura near Anaheim around the turn of the century, you knew about Holeshot. Maybe even got spanked by a customer of theirs. Credit that to the "do it once, do it right" approach that Aaron Bonk and partner Jon Spackman applied to every car going through their shop.

I heard about Holeshot while working as a contributor to *Sport Compact Car* and *Honda Tuning* magazines. At the time, we were hungry for engine swap articles and feature cars that reflected the Hybrid trend growing within the Honda street culture. Holeshot was the place many of those articles began. Over six years and hun-dreds of swap jobs, Bonk and Spackman developed and/or perfected many of the "standard" practices for Honda transplants still in use today. They also learned a lot about what NOT to do from the constant flow of botched and half finished swaps that they were hired to sort out.

Thinking about Holeshot's volume of work, I once asked Aaron what he did with the numerous unwanted D-series engines left over from all those swaps. Aaron shrugged. "We just push them out into the alley behind the shop and hope they'll disappear before morning."

Bonk's humble yet confident personality was mirrored in his black '95 Integra GSR. Clean and plain looking on the outside, there was little to give you the impression it was anything other than stock. But the turbo B18C under its hood put out 550 hp to the wheels. "Yeah," Aaron would admit, "it's pretty fast, I guess."

I can think of few more qualified to write "the book" on Honda engine swaps, and none more reputable. Enjoy.

E. John Thawley III
Automotive Journalist and Photographer

Acknowledgments

Special thanks goes to a few generous individuals for their unsurpassed wealth of general Honda know-how; Jon Spackman of Holeshot Racing for his assistance with the Honda electrical system and just about everything else; Brian Gillespie of Hasport Performance for his information regarding JDM engine codes and vehicle designations; Jacob Breyman and Billy Kelleman of Import Life, Eric Cardines of Honda Fiend, Robert Young, and George Hsieh and Dave Newman of HCP Engineering for their providing of vehicles, engines, and any and all swap components.

ENGINE SWAP CRASH COURSE

Automotive engine transplants are, without a doubt, anything but new. In fact, car enthusiasts and hot rodders have been swapping engines for the past 50 or so years. On the other hand, for those in the rapidly growing sport compact community, the engine swap is the latest, most popular trend. It's so preferred among Honda enthusiasts that they can single handedly take credit for more engine transplants than the rest of the sport compact community combined. It wasn't until the mid 1990s with the introduction of the third-generation Integra that the Honda engine swap craze had really taken off. Once the hardcore enthusiasts realized the chassis and engine similarities between the existing fifth-generation Civic and the latest Integra, the experimentation process began.

Why a Honda Engine Swap?

Unlike any other import or sport compact vehicle, the market for Honda and Acura engine swap products has grown beyond all expectations. There are a couple of main reasons for this, the most important of which is the bang for the buck. In fact, if you think about it, the whole import scene was built on that premise. You see, the average import enthusiast's age is somewhere between late teens and early 20s, so the chances of being able to afford a 'Vette or a Mustang Cobra are pretty slim. What they

need is performance on a budget. That's where the sport compacts and imports come in to play. A '91 Honda Civic for a couple grand is a bit more within grasp when on a typical minimum-wage budget. With all of the extra cash that is saved by going with a less-expensive car, the owner can now look into performance upgrades without breaking the bank. It isn't hard to see that it is those very same reasons that make the Honda engine swap industry so demanding. Not only are the vehicles themselves affordable, but you can literally double

your horsepower rating in many cases for less than the initial cost of the car.

In addition to its extreme low price tag, the second key point in favor of Honda engine swaps is the compatibility of parts. Many of Honda's vehicles share the same chassis, electrical system, and engine mounting points as some of their Acura counterparts. This means that in many cases, everything from motor mounts to computers will be totally interchangeable. This can be the making of a total no-brainer engine swap. Equally important, they'll also keep

Underneath the hood, this Integra is virtually identical to a '92–'95 Honda Civic, except for the engine. Although engine swaps are routinely performed on the older vehicles, the introduction of this late-model Integra is what really kicked things off.

Engine Swap Tips and Tricks

1. To free up room under the hood, try installing a high-performance fan from Hayden or Spal. These fans are significantly thinner than the Honda units and can be fastened to either side of the radiator as a push or pull unit.

2. Most of the engines in this book will require a minimum of 91-octane gasoline to be used at all times. Many of the JDM engines were designed for Japan's standard 93 octane.

3. It's wise to remove the throttle position sensor from the engine before installation. These sensors somehow always find a way of coming into contact with the firewall and breaking into several useless pieces.

This TPS was broken while at the wrecking yard. They usually snap right in half.

4. When looking for ways to reduce the vehicle's overall weight, don't overlook the air conditioning and power steering units.

Just look at all of the components that are involved in the A/C and power steering systems. If you're careful when you remove them, they can be quite valuable to someone else.

5. If a clutch is to be installed at the time of the swap, temporarily slide the disc onto the main shaft of the transmission before installation. This will confirm that the splines line up properly and will ensure (to the installer) that the correct clutch is being used.

6. Always use fuel-injection-style hose clamps. Standard hose clamps tend to put small cuts in the hose when tightened, which results in possible leakage.

Notice how these fuel-injection hose clamps don't have the ridges in them that you would normally find in a standard hose clamp.

7. Always use fuel-injection hose when replacing or adding fuel line. Standard fuel line found in most auto parts stores is not rated for the high pressure of a fuel-injection system and will burst after time.

8. When moving the engine into place with an engine hoist, try using an adjustable load-positioning bar so that it may be lowered at an angle.

9. Make sure that you understand engine serial numbers before selecting a donor engine.

10. Be careful when selecting a transmission and relying on serial numbers as the sticker often falls off.

11. Upon removing the speedometer cable on vehicles so equipped, be sure to put the small retaining clip in a safe place, as they are often lost in the cracks of the vehicle.

12. Instead of draining the air conditioning system and removing the compressor from the vehicle, tie it to the frame with some heavy-duty zip ties.

13. Be careful when selecting the proper OEM exhaust manifold. Many upper halves are not compatible with various lower halves. Just because they are both a B-series, for example, does not mean that they're compatible with one another.

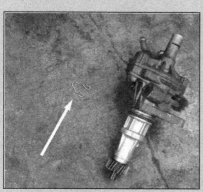

The clip pictured to the left will slip into the groove on the top of the vehicle speed sensor shown on the right. Without it, the cable will slip out of place.

With these heavy-duty zip ties in place, the A/C compressor is well out of the way for installing the new engine.

Notice where the downpipe connects to the upper portion of the exhaust manifold. Often times the bolt pattern differs between manifolds that you would assume would be similar.

The world's fastest Honda is the Progress Suspension Landspeed Civic. It relies on huge horsepower from a highly modified B-series engine. At over 200 mph, the stock D-series Civic engine just couldn't cut it. (Photo Courtesy of Progress Suspension)

your vehicle in line with many state regulations and emissions laws. States including California enforce strict regulations and guidelines on engines that you may or may not put into your vehicle. Fortunately for engine swappers, the fact that many of the Hondas and Acuras share the same chassis means most of these projects can be done within the confines of the law. Pretty cool, isn't it? You can double your horsepower, pass smog, and still maintain a factory appearance under your hood. No wonder the Honda is such a popular transplant candidate.

Decisions and More Decisions

So, now that you've decided to follow in the footsteps of the prevailing majority and go with a Honda engine swap, the next step is to settle on which

Several road-race teams rely on the larger and more powerful DOHC B-series engines despite their additional weight. Thanks to a wide variety of Honda suspension components available, handling can be improved upon regardless of the heavier drivetrain. (Photo Courtesy of Progress Suspension)

one. Keep in mind though, just because Hondas make up for most of the transplants out there, this doesn't mean that every Honda or Acura vehicle would be a good choice for a swap. In fact, with some vehicles it wouldn't even be a smart decision to modify them in any way, shape, or form at all. Let's cover some basic dos and don'ts, as well as some points that you might want to take into consideration before purchasing your next swap vehicle or donor engine. Some of these ideas have been acquired through lengthy research and others have just been stumbled upon accidentally. The fact that my business partner, Jon Spackman, and I performed a couple hundred or so engine swaps through our business, Holeshot Racing, doesn't hurt either. So pay attention and read on; chances are you'll find most of your questions answered here.

You'll want to take into account several aspects before purchasing the vehicle or the engine. One very important idea to keep in mind during this decision process is to be realistic. It is important that you be realistic with the vehicle you choose, with the engine you choose, and with the amount of money that you'll be shelling out. Most important, try not to set your sights too high and dive into a swap that could possibly take you six months to finish, or worse yet, get yourself into a project that will render your vehicle unstreetable when it's done. Don't laugh; scenarios such as these occur all of the time. Whether it's a hood that refuses to close or an oil pan that almost touches the ground, there

Unless you have the most pristine car on the road, it never really makes any sense to convert a carbureted car to fuel injection. With hundreds of pre-owned injected Hondas for sale, it's much easier to just start fresh.

are certainly engine swaps out there that shouldn't have been done. If you don't want to end up like one of these cases, pay attention and stick to one of the recommended engine swaps. The keys to a successful transplant project are careful planning, having an open mind to all of your options, and being practical.

Carbureted Downsides

When considering your options, you might want to heed a few simple suggestions. Now, I could ramble on and on about all of the engine swaps that I wouldn't do, but that might not be so productive. Even so, there are a few swaps in particular that customers and racers will inquire about most often and should be mentioned here with a word of caution. That being said, avoid doing a swap of any kind on any and all carbureted Hondas, including all vehicles manufactured before 1986 and many produced afterwards.

More than likely, you're planning on swapping in a fuel-injected engine. If so, converting the carbureted fuel system to fuel injection isn't an easy task at all. It consists of replacing the gas tank, fuel pump, fuel lines, and the entire electrical system. This is just unnecessary when the same chassis can usually be purchased in fuel-injection form for a little more money. The initial investment of buying a later-model car will far outweigh the consequences of trying to do the conversion. Every shop owner has that one customer who decides to go against what is recommended by the professionals, and I know that some of those folks are probably reading this book. To those people I say, good luck, because you're on your own. I just hope that you really love the underside of that '82 Civic because you're going to be spending quite a bit of quality time under there. The bottom line here is that in most cases, these carbureted-to-fuel-injection conversions simply aren't worth it.

No Need To Downgrade

Another Honda engine swap no-no is to do any type of transplant on a vehicle that is factory equipped with a six-cylinder. Contrary to what many of my customers think, you do not want to downgrade your Accord V-6 VTEC engine for an H22A Prelude VTEC engine.

Surprisingly, a couple folks have inquired about that one. Equally important, if you own a vehicle as nice as an Integra Type R or the '99–'00 Civic Si and you insist on doing an engine swap — don't. It might be a more sensible idea to start with the same car, but with a less expensive trim level, perhaps an Integra RS or a Civic DX. After all, most of the nice things that made the Type Rs and Si's cost so much are underneath the hood. Unless you've catastrophically annihilated your current engine, get rid of the car and get a lesser model that will more than likely be cheaper and lighter. Come on, there's no reason to let a perfectly good B18C5 or B16A2 go to waste.

Let's go over a couple of pretty unlikely scenarios and talk about what not to do. Let's suppose a certain manufacturer makes a motor mount kit that allows you to bolt an NSX engine into your '87 Prelude, rear-wheel drive and all. Sounds great, doesn't it? Well, before you go handing someone your hard-earned money for something like this, listen up. You want a word of advice? Just because you can do something or buy something doesn't mean that you should. Many prospective swappers are

Vehicles equipped with the six-cylinder engine should be left that way. Transplanting the four-cylinder engines covered in this book is an upgrade best suited for the lower-end models.

blinded by thinking that getting the engine to fit into the car is the swap in and of itself. Ask any good metal fabricator; with enough cutting and welding, you can pretty much put any engine into any vehicle. Ever seen a 5.0-liter V-8 in a Ford Festiva?

Seriously, you have many other considerations to worry about besides bolting the engine into place. For example, what seems to be the biggest nightmare for most folks is the electrical system. If the wiring seems to intimidate you a little, then you'll definitely want to check out Chapter 3. Other complications will arise from shifter mechanisms, fuel systems, driveshafts, and clutch components. The good news is that we'll cover in detail all of the Honda and Acura engine transplants that are recommended, along with tips, suggestions, and many recommendations. Read on, follow the instructions, and we hope you won't end up with an undriveable '87 Prelude with an Acura NSX engine shoehorned into place.

Gaining Weight

Now that you have a basic idea of what not to do, it's time to start narrowing down your choices of what you

should be doing. One very important aspect that will assist you in narrowing down your options is the weight of the new engine. You'll want to take into account the weight of your current engine and transmission and the weight of the new engine and transmission as well. In many cases, the heavier engine will produce some negative effects of understeer and in the most extreme cases may warp the vehicle's frame. If you don't know what understeer is, it's when the front end of the car pushes and doesn't turn as sharply as you want it to. If a little bit of understeer is a problem for you, then your decision just got easier; go with a lighter engine swap.

If understeer doesn't bother you, then you can move onto other criteria. For instance, you don't want to overlook the suspension. If you're putting a significantly heavier drivetrain into your vehicle, then plan on it sitting about 1/2 to 1 inch lower than before. Luckily, springs are available that are designed to withstand the additional weight. However, don't forget to work these costs into your budget.

If this is a problem for you, then you may want to look into an even lighter engine swap, or possibly modifying your current engine instead. Hey, engine swaps aren't for everyone. If these things seem like they might be an issue for you then you may want to look into turbocharging or installing nitrous oxide on your current engine. For the rest of you Hondaphiles who have the need for speed, read on. As far as weight and suspensions go, these points will be addressed in detail inside each transplant section covered throughout the book.

Being Realistic

If an engine swap is for you, then it's soon going to be time to choose the two most important pieces of the puzzle: the car and the motor. Once you've chosen a vehicle and you're caught up in the process of shopping for a donor engine, let's try not to forget that it is still a Honda. What I mean is, you want to stay somewhere within the realm of the Honda person's budget. After all, it is an economy car. Even so, there are absurd cases of individuals who will try and attempt the whole '87 Prelude/NSX thing. This won't only cost about five times more than what the car is worth, but the amount of cutting and fabrication that is involved will make it undriveable. Attempting a project like this will most likely defeat the entire purpose of going the Honda way. A word of advice: if you're considering doing a rear-wheel drive conversion or a six-cylinder swap, do yourself a favor and go buy a used NSX to begin with. You'll no doubt save yourself time, money, and headaches.

By the end of this book, you should be close deciding on the swap of your choice. At some point in time, you're going to have to ask yourself if you're capable of performing the work yourself. First off, you don't have to be a trained Honda technician to perform the swap, although it wouldn't hurt. However, you really should have some experience with the basic engine removal and installation processes. If you've never done this, you might want to consider having a professional perform your transplant, or get an experienced friend to help. Equally important, make sure that you have all of

Right-Hand Drive Conversions

The inside of this Integra didn't have to undergo any type of conversion at all. This is a genuine right-hand drive Integra Si-VTEC, direct from Japan.

In addition to transplanting high-performance Japanese engines into the USDM Honda chassis, many Hondaphiles choose to take it one step further. By swapping in the most coveted of overseas body parts and interior components, a true JDM vehicle can be made here in the United States. From installing a rarified JDM digital clock to a full front-end conversion, JDM components are sought after everywhere to get that special look.

Hardcore JDM freaks, including Dave Newman from HCP Engineering, won't settle for anything less than the real thing. Rather than building his own JDM wannabe, Newman opted to have an entire right-hand drive, JDM '95 Integra Si-VTEC shipped to his workplace. A true right-hand drive, daily driven vehicle, it doesn't get any more JDM than this. Even under the hood, several components are reversed.

Notice the brake booster, battery, A/C lines, and fuse box are all on the opposite side of the engine compartment.

This rare JDM front-end conversion is a popular addition to the DC Integra.

the proper tools, manuals, and parts before beginning the project. Missing any one of these elements is what turns a one-day swap into a one-month bus pass.

Don't even think about starting an engine swap project without one of these. A factory authorized service manual will be sure to answer any questions that you won't find here in this book.

Get yourself a service manual straight from the Honda or Acura dealership. The amount of extra information that is in the genuine Honda or Acura manuals as opposed to the local auto parts store manuals is well worth the extra cash. Keep in mind, you'll not only need the manual for your vehicle, but in most cases, you'll want a second manual for the engine that you will be installing as well. So, plan on buying one for each of them. For example, if you own a '94 Honda Civic Si and would like to do a '95 Acura Integra GSR transplant, then you'll need both manuals. Once you have your manuals, you can take an inventory of your tools according to what the books recommend. Air tools won't be absolutely necessary, although they do make life a lot easier when removing axle nuts and suspension bolts. However, a good old breaker bar will always get the job done with a little bit of extra elbow grease.

A word to the wise: If you do not have the right tools, don't start the project. Too many times over-anxious racers yanked the old engine out in a big hurry, only to find that a special tool that they needed was on backorder and not available until later that week. The same goes for all of your swap components. Make sure that you gather up all

Sure, air tools are nice, but not everyone has access to equipment like this in their garage. However, if you do, you'll find that the job will go much more smoothly and quickly than with hand tools alone.

Maybe this is the reason why many folks don't use air tools. The compressor can be just as expensive as the engine you're installing. Many rental yards rent these out for reasonable rates.

of your parts before even thinking about pulling out your old engine. This one has always puzzled me. I've seen countless individuals (even shops) start a project only to realize that the parts necessary to complete it aren't there. It wouldn't hurt to make a checklist and talk to others who have done this particular swap to make sure you have everything you need before you get started. If you look ahead, you'll see that all of the necessary parts and special tools will be covered in detail later on in this book.

No matter how you do it, the bottom line is to make sure that you cover all of your bases before getting started. If you feel that you aren't capable of the mechanical or electrical portion of the

swap, don't hesitate to take it to a professional or seek guidance. Sure, almost every car guy would love to do all of his own work, but sometimes it can be just downright dangerous if you don't know what you're doing.

To Wire or Not To Wire

Okay, so you're going to go ahead and do the swap yourself. It's hoped that you've decided on the donor engine at this point as well. Once you've acquired your engine and computer, it's time to tackle the wiring harness. Many folks will testify that this is among the most dreaded parts of the swap process. Wiring-harness modifications at many times can be very confusing for some and time consuming and tedious for all. Over the years at my facilities, many customers would do the mechanical portion themselves and tow the vehicle down for the wiring. Others would just bring in the entire harness out of the vehicle before they even started their swap. Both of these are excellent ideas and very commonplace. Several companies including Place Racing, HCP Engineering, and Hasport Performance offer custom wiring harnesses, so this option is now easier than ever.

For those true do-it-yourselfers, you'll want to make sure that you have those Honda service manuals handy that I mentioned earlier. You're definitely going to need them. Among other things, you'll want to make sure that you have lots of electrical tape, several butt connec-

The wiring portion of the engine swap scares some enthusiasts away from doing an engine transplant themselves. If wiring is not for you, there are many pre-made wiring harnesses available from the aftermarket.

Aftermarket Engine Mounts

When performing a motor swap that is going to require aftermarket engine mounts, sometimes the choices may seem overwhelming. With over a dozen companies manufacturing mounts and brackets for the '88–'91 B-series swap alone, it can be difficult to know if you are making the right decision. Do you want aluminum or mild steel? How about MIG welded or TIG welded joints? Maybe you want a billet mount instead. What grade of polyurethane is best to use? What about motor placement and axle-angle considerations? These are all valid questions and important points that should be answered before purchasing a set of custom engine mounts for any swap.

The most important consideration is motor placement. Is the oil pan going to sit too low, or worse yet, hit the ground? What about the cylinder head cover rubbing up against the hood, and how about the axle angle? All of these points will come into play when contemplating motor placement. Unfortunately, when dealing with engine swaps, there isn't one magic location to

These MIG-welded mild-steel engine mounts from Place Racing are less expensive than the billet-aluminum mounts.

install every engine without clearance issues. In many cases, it is a matter of comparing different engine placement locations and considering the pros and cons of each of them. If your car happens to be extremely low, then of course you'll want to consider whichever mount kit places the engine at the highest point. If broken axles happen to be the norm for you, then you will definitely want to go with whichever kit allows the least-aggravated axle angle. HCP Engineering, Hasport Performance, and Place Racing all manufacture mount kits that address each of these problems and place the motor in the most optimal position. It's safe to say that these companies have been in the business long enough to have addressed these issues long ago.

When choosing the material for a set of engine swap mounts, it's ultimately going to come down to three things: appearance, price, and strength. Although nothing beats the look of a polished, billet-aluminum mount, for some folks, being able to choose between many colors of mild-steel mounts is equally attractive.

This billet mount from HCP Engineering is the strongest available today.

Appearances aside, for durability, precision, and strength, nothing beats a Hasport Performance billet-aluminum engine mount. On your vehicle, if longevity, strength, and an exact fit are of the utmost importance to you, then you may want to consider investing in the more expensive billet-aluminum versions. This isn't to say that mild steel mounts are weak, but in comparison to a billet aluminum counterpart, they will fall short. As with any welded component, they can break and they do warp slightly due to the heat involved throughout the welding process. However, with properly engineered steel brackets being installed by the multitudes, warped and broken mild steel mounts from the likes of HCP Engineering and Place Racing are simply unheard of. Yes, billet aluminum is stronger, but you will have to ultimately decide if you actually need that much extra strength on a mere street car.

Regardless of which material you decide to go with, most engine mounts undergo a rigorous assembly process. Before any manufacturing, brackets and

Several pieces can be produced all at once thanks to CNC machines like this one at Place Racing.

mounts are designed using the latest CAD/CAM software, and a trial set is produced for test fitting. Once the prototypes are given the go ahead, production begins.

Engine mounts start out life as sheet metal and tubing, cut out by a CNC (computer numeric controlled) mill. With the pieces cut to size, further CNC machining takes place to drill, lathe, and bend various pieces into their final shape. Some steel mounts undergo a welding process with powder coating to follow. Billet-aluminum components will not require welding or coating. Once completed, special proprietary blends of polyurethane are formulated and poured into place. Finally, they are packaged, shipped, and ready for installation.

The pieces are MIG-welded together once all of the individual components have been CNC machined. Soon a special blend of polyurethane will be poured into these pieces of tubing.

This is just one of the many sophisticated pieces of engine-mount building machinery at Place Racing. Edward Luna operates the CNC brake for a future A/C bracket.

tors and ring terminals, a good set of wire crimpers, wire cutters, and a decent soldering iron. With these tools, the proper service manuals and a little electrical know how, you should be just fine concerning most swaps. Basic wiring guidelines will be covered in each swap section, but for the details, you're going to have to refer to the factory service manual.

Swap Components

It's now time to gather your parts together for the installation. The quality of the parts that you choose will reflect the ease of the work to be done, not to mention the final appearance. The first and most important group of parts that you're going to be gathering is the motor mounts.

Depending on what type of swap you'll be tackling, you're going to have three choices. Most swaps will require an aftermarket mount kit, which we'll discuss in detail later. Such aftermarket mount manufacturers include Hasport Performance, HCP Engineering, and Place Racing, just to name a few. Almost

as many swaps can be finished using a wide variety of OEM Honda and Acura mounts alone. The trick is finding out which ones you'll need.

Finally, there are a few swaps that will require you to custom fabricate your own mounts because there are none. All of the major mount selections will be covered later in the following chapters. However, keep in mind how fast the industry is changing, and motor mount kits are changing as well. You should research this aspect on your own before you begin the job.

Many other parts are just as critical to acquire beforehand, such as shifter mechanisms, driveshafts, suspension parts, throttle cables, hoses, and cooling fans. Of course, all of these parts won't be necessary for every transplant, but you'll have plenty to gather up before the big day. Follow the information in the swap sections and the chances of a successful engine transplant will be greatly improved.

Now that you have a little bit of background on the Honda engine swap scene as well as some guidelines for

choosing a swap, it's time for you to dig a little deeper. The rest of this book is going to cover the most popular engine transplants for the last several years. Browse through the swap chapters, as they're in no special order, and by the end of the book you should have a pretty good idea of what swap you'd like to do, as well as enough information to put you on the right track.

Up until just a few years ago, if you wanted an engine swap, you had to custom fabricate your own mounts, like this one. Luckily, several manufacturers produce engine-mount kits for almost any Honda engine swap.

How to Use This Book

You'll find that once you've gone through the first few chapters of this book you may want to jump around between the different engine swaps that interest you. That isn't a good idea because certain topics that pertain to all swaps are only mentioned in some places – there's no way there would have been room to explain it all over again every time. This book is designed as a guideline to help you figure out what may or may not be a good engine transplant for your vehicle. It makes every attempt to give you all of the information you'll need. However, with so many aspects of the engine swap and a limited number of pages, not every little detail was covered. With help from a factory Honda or Acura service manual, all of your questions should be answered. It is assumed that much of the information found in these

manuals is common knowledge among readers. Below are a few points to consider before going any further.

1. Although not every possible engine swap is listed here, most engine swap pros agree that these are among the most popular being performed.

2. When reading about possible donor engines for your vehicle, don't be discouraged if you don't see the exact B series or H series that you were hoping for. Refer to the engine swap chart at the end, as there is not enough space in every chapter to list every single engine/vehicle combination.

3. Although automatic-equipped vehicles may certainly be equipped with a manual transmission, it will be assumed that vehi-

cles referred to in this book have manual transmissions unless otherwise stated.

4. You must be absolutely sure to check your local emissions laws and regulations before selecting a donor engine. Just because an engine swap is mentioned in this book, do not assume that it is necessarily legal in all cases.

5. Last, remember that just because there is another way to do something doesn't mean that it's right. And just because you heard that your brother's friend's girlfriend's brother has done it one way, doesn't necessarily make it right, either. What you may consider is that the techniques and procedures on the following pages are generally accepted by most professionals, and are considered standard practice.

SWAP SAFETY

When doing any type of automotive work, it's always important to make safety a top priority. Using the proper equipment and following instructions can mean the difference between you lying in a hospital bed or putting the finishing touches on your engine swap. Safety in the workplace cannot be stressed enough, so the subject will be dealt with before we go any further.

Clean and Uncluttered

Before starting any project you'll want to make sure that your work area is ready for you. In addition to having plenty of lighting and ventilation, there are a couple of other key points to remember. An environment that is clean and uncluttered is much easier to work in. This is an excellent habit to get into, especially if you find yourself working in your garage at home. It will be a lot easier to find what you're looking for and reduce the risk of hazards. How many times have you tripped over a pile of tools left on the ground or slipped in a puddle of oil that you swore that you'd clean up later?

Before you start your project, make sure you have a clear area set aside to

work in and invest in lots of towels for the inevitable oil and coolant spills.

Tools and Equipment

Once you've set aside a nice place to do your swap and obtained plenty of shop towels, it's time to sort through your toolbox and make sure that you are prepared. We hope your collection of tools consists of Snap On, Craftsman, Matco, or any other high-quality American manufacturer. If so, then you'll be in good shape. If you open your box and find a ratchet that you picked up from the local 99-cent store, then it's time to go tool shopping. It's just plain silly to go the cheap route when purchasing tools. Buying quality tools to begin with will mean a larger initial investment, but the longevity and durability will far outweigh the costs in the long run. Besides, most of the more expensive brands come with lifetime warranties. You won't get that with your 10-dollar socket set.

A hydraulic floor jack of some sort is necessary if you aren't using a lift. Be sure to get one that's rated for at least the amount of weight you'll be lifting.

Your shop equipment is as important as the tools you'll be using. Make sure that the hydraulic jack that you're using is suited for the weight that you'll be lifting. Moreover, remember that the jack should always be accompanied by a suitable jack stand on each side of the vehicle at the appropriate lift points. Consult your factory service manual for the exact locations.

If your backyard doesn't have an automotive lift in it, you're going to need some type of engine hoist. Most rental yards will let you take one for a day for significantly less money than purchasing one outright. Same thing goes for engine hoists as the tools: don't go the cheap route. When borrowing or purchasing a used hoist, it's always a good idea to look it over thoroughly

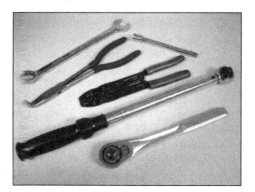

These tools from Craftsman come with a lifetime warranty, should you ever need it. Nothing's worse than a knuckle buster when wrenching on your car. Invest in a good set of tools to avoid any unfortunate accidents.

Always use a couple of jack stands underneath the frame rails to go along with your floor jack when you're working under a car. You never know when the floor jack might fail or slip out of place.

When you're disconnecting fuel lines, keep in mind that they can spray quite a distance if the system is still under pressure. Wearing some sort of eye protection is always recommended when you're working on the fuel system.

before hanging a 400-pound, toe-busting engine off it.

Make sure that the wheels spin freely and that the hoist is easily maneuverable in your workplace. Check the nuts and bolts on the hoist for tightness. If you're unfamiliar with operating any of this equipment, find yourself an instruction manual or someone who can help you before beginning the swap.

Dangerous Fluids

Once you begin the transplant process, there are other key safety points to keep in mind. You'll want to make sure that you have access to running water nearby in case you get something in your eyes. Lying under a vehicle on your back, you'll almost inevitably find some type of fluid on your face at least once. No matter what it is, you'll want to wash it off immediately or face the consequences of burning or irritation. Another source of hazardous fluids is the fuel system. The automotive fuel system is under about 50 psi of pressure. Even after the car is turned off, some of that pressure remains. When initially disconnecting fuel lines and hoses, make

sure you adequately relieve the fuel pressure according to the factory service manual. Have a couple of rags handy to clean up any spills when working around fuel lines, and of course, no smoking. A fire extinguisher is also something that you should always have nearby when doing automotive work.

Following these safety guidelines won't only protect you from unnecessary harm, but it will also play a significant role in speeding up the swap process. A well-lit, uncluttered environment with the proper tools and quality equipment will ensure a successful engine transplant.

If you're installing the new engine through the top of the engine bay, you're going to need a cherry picker. Double-check the load rating before hoisting anything up with it.

As much as most car guys like to do everything themselves, sometimes they need help. Don't be afraid to ask for help if you don't know what you're doing; find somebody who does. Besides, it always makes it easier if you have a helping hand.

ELECTRICAL ESSENTIALS

Most folks would agree that the electrical system could prove to be the most confusing part of any engine swap. If you talk to anybody who has done his (or her) own swap, most will tell you that the wiring gave them the most trouble. This is true, as in most cases it does require intermediate to advanced automotive skills, as well as a solid comprehension of wiring diagrams and schematics. Remember those service manuals mentioned earlier? This is the part of the swap where you'll need to open those up to the electrical section. Winging it won't get you too far when it comes to wiring harnesses.

Unfortunately, for those of you who plan on installing an early model B16A, ZC, or any other "JDM only" engines, unless you're fluent in Japanese, you aren't going to be provided with any wiring diagrams from Honda of America. There is some good news, however. This information is available on various Internet forums and chat rooms, as well as certain Honda information sites.

Electronic control units (ECU) not only operate the vehicle's electronics and emissions systems, they also store valuable information. When problems arise, many answers can be found thanks to trouble codes stored in the computer.

However, if you don't have Internet access or just don't want to mess with any of the wiring, then a premade harness from Hasport Performance, HCP Engineering, or Place Racing will be in order. If wiring isn't your field of expertise, then this might be just the route for you.

OBD Explained

Speaking of electrical systems, one of the most misunderstood elements of the engine swap process is the OBD (on board diagnostics). First implemented by General Motors in 1981, on-board diagnostics was eventually mandated to be in place on all vehicles sold in California by 1986. OBD's ability to oversee the vehicle's emissions components worked so well that it became standard issue on all vehicles sold in the United States. In addition to being a watchdog regarding the emissions system, OBD was also designed to inform the driver of a malfunction, as well as to record the data in the ECU (electronic control unit) for retrieval by a technician.

As the result of several shortcomings in the original OBD system, the Clean Air act of 1990 passed and required a redesigned OBD system by 1996. Although Honda introduced changes to its OBD system on the '92–'95 models, what we refer to as OBD I is more similar to the OBD 0 system than the later OBD II. In some states, smog tests differ slightly between OBD II vehicles and earlier models.

Those with the OBD II system will, in most cases, undergo a two-part emissions test to pass. The first part involves hooking up a testing device to the data link connector of the vehicle's computer. This process will allow the technician to recall any emissions problems that are present or that may have occurred in the past. It's important to be aware of this testing procedure before the engine swap. The second part of the test simply involves the fuel filler cap and affirming that it is making a positive seal. Other than that, the smog test is much the same for OBD II and earlier models. It's important to keep in mind that OBD II equipped vehicles will be scrutinized much more than OBD 0 and OBD I vehicles come smog time. Engine swaps performed on these chassis will require more thought to select the proper computer and emissions components in order to be legal. It's always best to check with the proper authorities prior to shelling out the cash on any engine or swap parts.

OBD and Computers

Okay, so now that you know what OBD is, you'll need to know why you

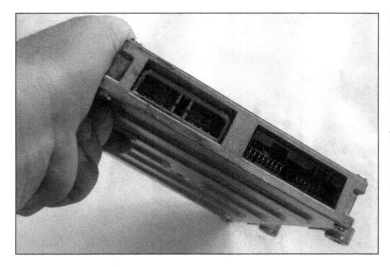

Pictured here is an OBD 0 Honda ECU. The easiest way to identify the OBD type of a computer is by first looking at the plugs. OBD 0, OBD I, and OBD II computers all feature different types of connectors.

need to be concerned with it. The proper computer or ECU selection is going to be a fundamental element of a successful swap. First, you need to understand that you'll encounter three different types of Honda ECUs. The first is the OBD 0. These can be found on most '91 or previous fuel-injected Honda or Acura cars ('90–'91 Accords are OBD I, though).

The second is the OBD ECU, otherwise known as OBD I. These are commonly found in most '92–'95 Hondas and Acuras. The third type is OBD II, which you'll find in everything from the '96 models on up. It is important to note that due to the variance in the OBD systems, the electrical connectors and pins are different on all three of these models and aren't easily interchangeable. Not only that, but they'll only recognize cer-

tain types of sensors and fuel injectors that are compatible only with the corresponding OBD system. In most cases, it is best to stick with whichever OBD type that your vehicle originally was equipped with. Ideally this will leave you with the least amount of wiring to perform and will keep you in line with the state emissions.

However, if you do have the right combination of components and your skill level in automobile electronics is very high, then almost any ECU discussed in this book can be interchanged into any vehicle. You'll find that in some cases this will be unavoidable. In many scenarios, you won't have an option to run an OBD 0 ECU because it simply doesn't exist, and you'll have to resort to some fairly involved wiring procedures.

Some of the later-model swaps won't permit you to use an OBD II ECU because of Honda's built-in anti-theft system. This puts great importance on selecting the proper sensors to go with the ECU. Choosing the right sensors and ECU will be covered in detail for each swap, since each transplant will differ slightly.

Choosing the Computer

When selecting a computer for your swap, several criteria must be answered before the purchase. The first and most obvious pieces of information that you'll need to have at the time of ECU selection, or any other swap part for that matter, is the year of the vehicle and the year of the donor engine. Matching the OBD level of the engine and computer to that of your vehicle would yield the easiest wiring. However, this won't always work in your favor (more on this in the following chapters).

The second point to consider when selecting your ECU is the transmission you'll be using. You'll need to decide between an automatic or manual gearbox. If you plan on running an automatic transmission then you're going to need an automatic-style computer, and if you plan on using the manual transmission, you'll need a manual style computer. When using an automatic, you also need a separate computer for the transmission to function properly on certain engine swaps.

Although all Hondas manufactured before 1992 are equipped with OBD 0, a few had OBD I implemented ahead of schedule. Select '91 Preludes and all '90–'91 Accords received OBD I.

This Integra crankshaft position sensor is one of many OBD II-only sensors that were implemented in 1996. Using a USDM OBD II ECU without this sensor will definitely throw a malfunction indicator (check engine) light.

Selecting the proper computer for the engine transplant can sometimes be tricky. Once you've decided on what you need, simply check the serial number on the side of the unit to make sure you're getting what you want. The P28 in the center indicates that this is from a '92–'95 Civic Si or EX.

OBD Conversion Harnesses

With Honda and Acura engine swaps ranging from vehicles as early as the mid 1980s to 2003, the interchangeability of parts can at times seem very enticing. This is just as evident in the electrical system as it is concerning any other aspect of the swap. With the introduction of OBD I, and then OBD II, the swapping of computers and wiring harnesses has become more and more realistic for the average home mechanic or do-it-yourselfer. However, this hasn't always been the case. In the past, if you wanted to use an Integra Type R ECU (which is OBD II) for example, in your '93 Civic (OBD I), then you were in for a weekend of wiring. Many enthusiasts used the wrong computer as a way to avoid this extra work, and the results were less than desirable.

Fortunately, the aftermarket has once again come to the rescue. The product that has made such an impact on the swap scene is the adapter harness. Companies such as Hasport Performance and HCP Engineering produce adapter harnesses that allow quick interchangeability of ECUs of different OBD levels. What once took hours, or even days of wiring, can now be done with an adapter harness consisting of the appropriate male and female connectors on each end.

In the old days, if you wanted to change the OBD status of your vehicle then you would need to spend some time underneath your dash. You would need to cut the connectors off the underdash harness that would normally plug into the computer and replace them with the proper connectors for the new ECU. One of the most difficult aspects of this process doesn't even involve the wiring. Perhaps the hardest part is locating a junkyard that will let you cut the plugs off one of their cars. Many yards are unwilling to destroy an entire underdash harness for the sake of making 20 bucks.

If you are lucky enough to acquire these connectors, then the real work begins. Using the service manual for your vehicle and your computer, reconnect all the wires to the appropriate pin locations on the new plugs. Things can get difficult for the novice, since some plugs don't have locations for some wires, and some plugs have two locations for the same wire. With over 90 percent of the wiring underneath the dash, it's easy to see why this isn't a fun job.

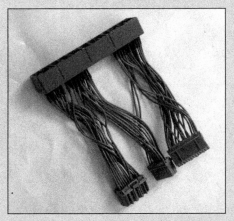

Without an adapter harness like this one from HCP Engineering, every single one of these wires would need to be added to the vehicle harness one by one.

Although OBD conversion harnesses are not available for all vehicle and computer combinations, quite a few are being offered and new ones being developed as we speak. Because OBD conversions are not legal in all states, it's best to check with the local authorities first.

A third key point deals with possible upgrades that your engine might go through. It's important to decide whether you'll want to have your computer reprogrammed or chipped in the future. Many automotive chip manufacturers are unable to reprogram OBD II computers; some can, but it is usually very expensive. A cheaper alternative is to start out with an upgradeable computer initially so that when you do decide to have it chipped, it won't be as costly. OBD 0 and OBD I ECUs are both easily reprogrammable.

Last but not least is the concern over smog legality. Most state laws require the donor engine to be the same year as the vehicle or newer. This not only includes the engine itself, but all of the electrical components, sensors, and most importantly, the computer. So, what do you do if you live in a highly enforced low-emissions state, and it just isn't possible to use the proper computer? Well, this is hardly an uncommon situation among enthusiasts in California (the strictest emissions controlled state). The reality is that there are innumerable illegal engine swaps done every year. These cars are illegal in that they might have the wrong computer, possibly a mismatched intake manifold, or maybe an outdated sensor for that particular ECU. In many cases, the tailpipe emissions that these vehicles put out is

Some Honda and Acura automatic transmissions rely on a separate computer that controls the shift points. When swapping in a different auto trans, be sure to get yourself the accompanying computer for it as well.

You'll find that most engine swap vehicles will burn just as clean as those retaining their original engines. Unfortunately, this is just one part of the smog check process. You may also need to pass the visual inspection.

Gauges and Instruments

After certain engine swaps, you may find that your current gauges simply won't provide you with the information that you need. Whether they don't register the miles per hour that you desire or if they're just missing a tachometer, often the factory gauge cluster has to go.

There are a couple of solutions to this problem. For folks in search of the most factory-looking appearance, in many cases an OEM gauge cluster may be retrofitted into place very easily. When going this route, the first thing to do is look for other models of your vehicle and see what they offer. If that doesn't work, often the JDM models will have what you are looking for. If all else fails, in a few rare cases a cluster from an entirely different vehicle may fit snugly into your dashboard. Among the most common gauge-cluster conversions, many folks with base-model Civics install the top-of-the-line clusters by simply plugging them in. Folks with '88–'91 Civics often use the '88–'91 Si cluster, or the JDM cluster that was mated with the ZC engine. Owners of '92–'95 Civics will find a suitable cluster in any '92–'95 USDM Civic or JDM SiR models. A popular choice for '96–'00 Civic owners is the '99–'00 Si cluster. Several other vehicles and gauge clusters

may be matched to one another. Chances are, if the vehicles are similar, the cluster will slide right into place.

For those folks who just aren't able to find what they're looking for, sometimes turning to the aftermarket is the only option. Companies including AutoMeter make a wide variety of gauges to meet all your metering needs and more. Of course, these won't provide you with the factory-looking appearance that an OEM cluster

will, but on the other hand, AutoMeter gauges do have a cool racing-inspired look to them. Between all the tachometers, speedometers, and temperature gauges offered, you should easily be able to find what you're looking for. Perhaps the most difficult part is the installation. Since each gauge needs to be hooked up both electrically and mechanically, sometimes custom gauges prove to be a much more time-consuming route.

This JDM gauge cluster is a perfect post engine swap addition. Notice that the tach reads up to 10,000 rpm and that the speedometer reads in km/hr.

When using aftermarket gauges like these from AutoMeter, you can get as creative as you want with the installation.

no more than that of a perfectly legal California-compliant swap, but the problem stems from a visual standpoint.

Of course, I can't condone illegal engine transplants, so my advice is to make every possible attempt to make your swap as legal as possible in order to avoid long-term problems for yourself.

Wiring Harnesses

Computers and sensors alone aren't the only components involved in the electrical system. Almost equally important is the wiring harness. No matter what car is chosen to receive the engine, it will in no doubt have some type of wiring harness under the hood, as well as another harness of some sort underneath the dash. From now on, these will be referred to respectively as the engine harness and the underdash harness. You'll find that, in most cases, the underdash harness will remain intact,

Using the original engine wiring harness of your vehicle when doing an engine swap will save you quite a bit of work. Once you get properly hooked on to the new engine, you'll find that it will plug right back into your vehicle.

yet very seldom will you abstain from modifying the engine harness in some way (be it large or small). It's important to note here that it's possible to use the original engine harness that came in the car in every swap mentioned in this book. Although several USDM donor engine harnesses may be fully compatible with the new vehicle, they usually aren't.

Using the Right Harness

Many are misled into believing that they have to use the harness from the donor engine; this is usually not the case. Aside from various needed electrical connectors, most of the donor harness will soon end up in the trash. There are a few reasons for disposing of the new engine's harness and using the old engine's harness. Before explaining that, a couple of facts need to be understood

about how Honda lays out the vehicle's wiring. In all cases, the engine harness must connect to the underdash harness at some point in order for the electrical system to function. In certain cases, this occurs at one or two places on the firewall, and in other cases, the connection is made beneath the glove box with a different connection.

By using the original harness that came with the vehicle, you'll eliminate any work involved in connecting the underdash harness to the engine harness. If you purchase a JDM engine, which many of you will, then the engine harness equipped on that engine will be flipped around due to the vehicle being right-hand drive. In order to use this harness, it must be cut off at the point where it connects to the underdash harness and lengthened. The harness won't fit the

way Honda intended, causing a sloppy look and several hours of unnecessary labor. In most cases you'll also find that the harnesses found on junkyard engines are usually cut, stripped, or thrashed.

There are a few criteria to look for when deciding if you can use the engine harness off of a junkyard engine. First, the donor engine must be from a USDM vehicle that was standard left-hand drive. Second, it goes without saying to make sure that none of the major connectors are chopped off or missing. Third, the harness must usually be of the same OBD type as the vehicle in which the engine is being swapped into. It doesn't necessarily need to match the computer being used. Last, you'll only be able to interchange harnesses of vehicles that share a similar chassis (such as Civics and Integras). If your prospect engine harness meets these criteria, then you can in many cases save yourself quite a bit of work. Keep in mind though, that in some of the more involved swaps, this will just not be possible. For example, no matter how hard you try, putting a Prelude engine harness into a Civic will never make sense.

trying to finagle a JDM-style engine harness to a USDM vehicle, you'll find it'll *hours of unnecessary labor. Since the JDM harness is designed for right-hand* *harness will be almost a mirror image of its USDM counterpart.*

Wiring Harness Modifications

At this point, you're probably wondering if you need to modify your wiring harness, and what exactly that entails. The first thing to understand is that when you're replacing an engine with a different one, you're inevitably going to be adding, changing, or eliminating sensors, not to mention computers. Although you'll usually be using the original engine harness found in the swap vehicle, you'll find that trying to plug this engine harness onto the new engine won't happen easily. At the very least, wires will have to be lengthened and plastic electrical connectors will have to be swapped.

In some of the more difficult situations, you'll find yourself having to add additional wires inside for sensors that weren't present on the old engine. These wires need to be connected to the ECU in order to function properly. Pay attention to which sensors are shared between the original engine and the new engine, and which ones are different. These newfound sensors must be recognized and located in the service manual. By following the wiring schematics, the sensor can be wired to communicate properly with the computer. In several situations, this can be accomplished by using an existing wire in the harness that is no longer needed, or simply by running wires directly from the sensor to the ECU. Of course, running the wires inside the harness will look much nicer. You'll need the plastic electrical connectors from the discarded engine harness in order to hook up the wires to the new sensors. This will provide you with the appearance of a factory finish.

Electrical Tools

Most of the tools you'll need to perform the modifications to the wire harness are probably already in your toolbox. Based on preference, you may choose to use standard crimp-style electrical connectors or the more permanent method of soldering. Both have their advantages, but if you're looking for that factory, permanent look, then nothing beats soldering. Whichever method you choose, you'll also need an ample amount of electrical tape, a razor knife, plenty of various colored wire, and a decent pair of wire strippers.

Electrical work is tedious and time consuming; remember to be patient and work carefully.

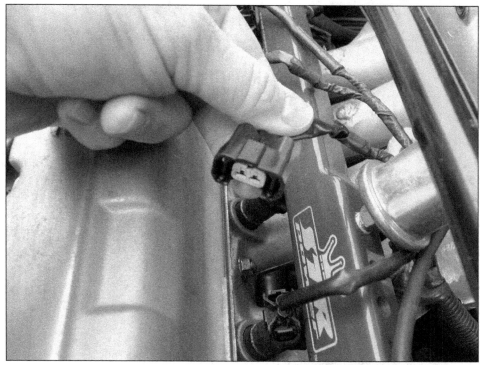

More than likely, any engine swap is going to entail at least some minor electrical work. At the very least you'll find that you'll need to swap a connector or two to accommodate a different style sensor.

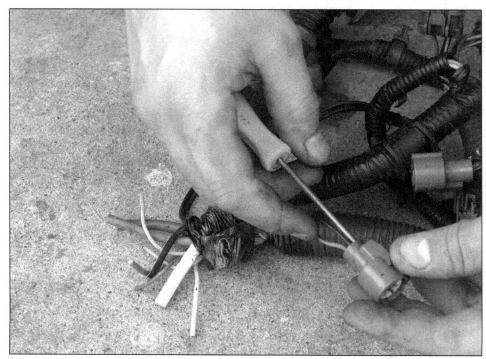

When swapping electrical connectors, often the pin can be removed from inside the plug rather than cutting the wire. If you don't have the proper pin removal tool then a small flat-head screwdriver will suffice.

1984 TO 1987 CIVIC AND 1986 TO 1989 INTEGRA

The useless SOHC engine is being hoisted out of this first-generation CRX to be replaced with a larger and more powerful B series.

These first-generation Integras are similar to the Civics in many ways. This makes for an identical engine swap scenario.

When the '84 Civic and CRX were introduced, it would be fair to say that automotive enthusiasts were not in a rush to get their hands on them. The Honda performance craze, as we know it, is still almost a decade away. What's interesting though, is how this '84 Civic has made a sort of imprint on the world of engine swaps.

The '84–'87 Civic and CRX are often overlooked as candidates to receive engine swaps. With a curb weight lower than any other vehicle in this book, and a price tag that's equally low, these are perhaps the main characteristics that have kept the AH chassis Civic and AD chassis CRX in the swap domain for so long. You'll find that the '86–'89 Acura Integra (DA chassis) isn't much different. Weighing in the neighborhood of 2,300 pounds, these are the lightest of all Acuras. The first Integra ever produced, and the first Honda vehicle in the United States to receive a dual overhead camshaft engine, the DA chassis entered some pretty stiff competition from the likes of the Celica and Eclipse.

Sibling Compatibility

One of the main issues that prohibits the third-generation Civic and first-generation Integra from being as popular as the newer models is their lack of aftermarket support. Since few high-performance suspension and braking components are available, upgrades after the engine swap can be difficult. However, some suspension and brake components from the Integra will fit on the Civic. The Integra's front knuckle and wheel hub assembly is compatible with the Civic, making it a breeze to install the larger front brakes of the Integra. As for the Integra, further braking upgrades include installing a pair of slotted rotors and some high-friction pads. Manufacturers such as Tokico offer upgraded suspension components including high-performance shocks for both of these cars.

Thanks to the compatibility between these two vehicles, suspension components may be handed down to the Civic from the Integra. You'll find that suspension and brake upgrades like these are extremely important when doing an engine swap.

'84–'87 Civic and CRX Offerings

The '84–'87 chassis Honda Civic was originally offered in several different forms and trim levels. The Civic may be purchased in the form of a three-door hatchback, a four-door sedan, or the rare wagon model. Trim levels consisted of the Si, DX, GL, and S models. Wagons were only available as DX, WV, and 4WD models. Another variation of the third-generation Civic is the CRX hatchback, which was first introduced in 1984. The CRX is available in Si, DX, and HF models.

The lightest model of all, the CRX, is a perfect candidate for an engine swap. It's rare that you'll find one as immaculate as this '87 Si.

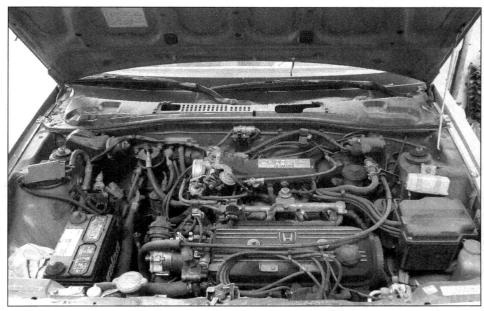

Remember it the way it was; this one's on its way out. These EW engines are absolutely worthless as far as performance is concerned. A B-series engine will soon find its way underneath the hood.

Looking under the hood of any of these Civics shows you why the aftermarket has never jumped on them. The top-of-the-line '87 Si model's D15A3 engine produced only 91 horsepower at 5,500 rpm, leaving enthusiasts few options regarding horsepower. Other offerings included the older Si's EW4015, the other lowly models' D15A2, D13A2, EW3025, EV1035, EW1035, EV1025, and EW1025. These gutless engines are barely worth the amount of money you'd get from recycling them, so it isn't necessary to waste any more time with them.

'86–'89 Integra Offerings

As far as the Integra, there certainly weren't as many variations to choose from. The Integra was originally offered in only two body styles. The four-door and two-door versions could both be had in RS and LS trim levels. Every first-generation Integra received the fuel-injected D16A1 and D16A3 ZC engines; varying interior amenities and a sliding moon roof are all that differentiated the RS and LS.

OBD Issues

All of the '84–'87 Civics and '86–'89 Integras are OBD 0. In fact, OBD I was

not fully implemented until 1992. For the sake of simplicity, we'll only concern ourselves with fuel-injected cars. Injected models include the '86–'87 Civic Si, the '85–'87 CRX Si, and of course all the Integra models. As far as the twin-cam engine swaps in this book are concerned, these are your only options. Of course, carbureted models may be used, but then, why bother?

When going for the CRX as opposed to the standard Civic, it had better look like this from the back; otherwise, you're in for some serious work. The Si is the only CRX that is equipped with fuel injection.

OBD 0 Swaps

Of the engine candidates for these swaps, the easiest from an electrical standpoint would be an engine that

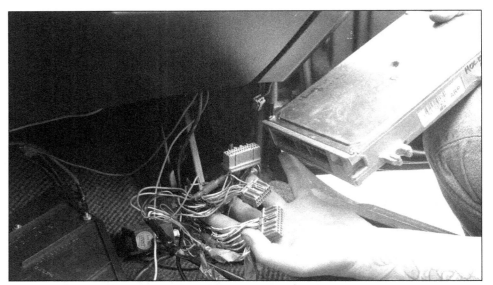

OBD wiring conversions are time consuming and tedious. You need to cut off each one of these wires entering the ECU of this late-model Integra and solder them into place on the Civic or CRX of choice.

came with OBD 0. Being able to plug the new ECU directly into the original wiring will help lead to a pain-free wiring process. Although there are some electrical issues that do need to be tackled, choosing an OBD 0 ECU will eliminate a major portion of the work. Unless an engine is only available with a later OBD form, OBD 0 swaps are the ultimate option for these vehicles.

OBD I and OBD II Swaps

What about those folks who would like to stuff a late-model Type R engine under the hood, or any other OBD I or II engine for that matter? As much as I'd like to recommend choosing a different project vehicle, this isn't always what the customer likes to hear. Actually, with the right amount of auto electrical know-how, any OBD-type ECU may be wired up into place if need be. USDM OBD II conversions should be avoided however, as they require the addition of several smog components, including a bung that must be welded onto the gas tank for an emission sensor. Needless to say, the JDM units are a much better choice if OBD II is going to be used.

If you aren't up for a major wiring conversion but still insist on a late-model engine, backdating the sensors and computer is always an option. By swapping several older components onto the new engine, including the distributor, oxygen sensor, and fuel injectors, OBD 0 status may be retained.

For those of you who still insist on converting to a later OBD ECU, further wiring modifications must be made inside the vehicle. Depending on which computer is used, the correct connectors may need to be sourced from a wrecking yard vehicle. These can simply be cut off near the ECU that you plan to use. Remember to leave yourself several inches of wire for connecting the new ECU plugs to your existing harness.

The old ECU in the Civic or Integra may be located underneath the passenger seat. Once the ECU is removed, the original plugs must be cut off, and the new plugs must then be soldered onto the vehicle harness of your Civic or the Integra. Then you'll be able to plug in the new ECU. Newer OBD computers display trouble codes on the gauge cluster, instead of on the face of the computer as with the older models. The appropriate wiring must be added to the ECU so that the codes can be displayed on the dash when necessary.

Service manuals for both the vehicle of choice and the ECU must be consulted when connecting the wiring between your car and the new computer.

Until a plug-in type of adapter harness is available for this chassis, a major wiring job is inevitable.

Wiring Harness Issues

In all cases, regardless of which engine or computer combination is to be used, the original Civic or Integra engine harness must always be reused. Harnesses off the donor engine must be saved for special plugs and connectors, but they will eventually end up in the trashcan since they're not compatible with the vehicle's connectors. In most cases, the engine wiring harness will need to be lengthened in various locations to accommodate the new engine. Plugging the injector plugs of the harness into the injectors on the new engine will give a good starting point to see which plugs must be cut and lengthened. When cutting connectors and adding lengths of wire, make sure the right wires go to the right spots in the plug.

ZC Engine Swap

The ZC is definitely among the easiest DOHC engines to transplant into this Civic and CRX chassis. These engines were standard equipment in JDM Civics of this era. We won't be covering the first generation Integra here, of course, as it is already equipped with the USDM version of this engine.

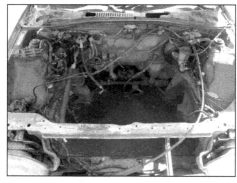

In addition to the fact that this CRX is left-hand drive, few differences exist between this chassis and the JDM version equipped with the ZC engine.

Since the Civic and the Integra share the same engine compartment, this installation process is nothing short of a bolt-in procedure. For this reason, the

ZC engine transplant should be considered one of the simplest engine swaps in this book. As with most any ZC swap, typically one to two days is more than enough time to get up and running. Whether the installer is an inexperienced novice, or a highly trained Honda swap technician, this transplant should be neither time consuming or difficult.

ZC Engines and Transmissions

Of the many ZC engines available from the wrecking yards, there are several sensible candidates to choose from. The D16A1 and D16A3 engines found in the '86–'89 Integra, although not labeled as ZC engines, are really USDM versions of the ZC. The two different stampings (A1/A3) indicate minor changes that were performed with the production of the '88 model. You can find JDM ZCs in the '86-'89 Integra GSi and RSi models, which are stamped with a ZC. Although most of these engines have not been manufactured since the late 1980s, like most ZC-style engines, they're still relatively easy to find in wrecking yards across the country. Other ZC engines came in the JDM '85-'87 Civic and CRX Si. You can easily identify these engines by their brown (instead of black) valve cover.

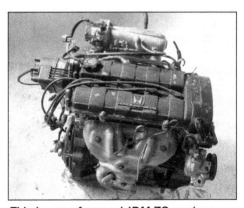

This is one of several JDM ZC engines that you may encounter at your local engine yard. Pay attention to the location of the driver-side engine mounts to be sure you're getting the proper one. Make sure it's on the front, not on the side.

Be certain when choosing any of these engines that it is a fuel-injected model, as there are many carbureted versions floating around. When shopping

Since all ZC engines have provisions for an intermediate shaft, you'll need to make sure you pick one up from the junkyard when you bring the engine home. The 3 empty bolt holes that you can see around the bottom left of the blue oil filter are for the halfshaft.

for a ZC engine with the standard black valve cover, it can be relatively easy to get the wrong one. Several ZC engines with black valve covers were produced, so it's important to be able to tell them apart from one another. An easy way to identify the proper engine for this swap is by verifying the engine mount locations. The key to identifying any of the ZCs mentioned in this section is to make sure that the driver-side engine mount is located on the front of the engine. This is clearly different from how most driver-side mounts are recessed inside the timing belt cover on the side of the engine. Get the wrong one and you can forget about bolting this engine into place on this car.

When selecting a transmission, it's important to take into consideration the differences between the Integra-style ZCs and the ZCs that are used in Chapter 5. Although both are ZCs, they have very different dimensions. The same thing applies to the transmissions. Just make sure you're using either the '86–'89 USDM Integra or JDM Integra-style ZC transmission. Either of these will bolt up into the Civic or CRX chassis as if it was meant to be there. Unlike other ZC swaps that you'll encounter in this book, the Civic and ZC transmissions aren't interchangeable on this particular project. It's always best to make sure that the transmission that you want to use is attached to the engine at the junkyard, and to take them both home at the same time. When split in two, sometimes one

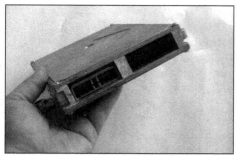

Most engines sold at a salvage yard will come with the proper ECU. If not, be sure to select the proper OBD 0 computer, as it will be necessary for the ZC engine to run properly.

or the other can get misplaced or sold as the wrong unit, and you might end up with the wrong transmission.

ECUs and Wiring for the ZC

Before doing any type of wiring, a computer must be selected. There are really only two computers that'll plug right in and work. These ECUs are the PG7 (USDM) and PM7 (JDM) units, which can be found in the '86-'89 Integras. Be sure to identify these serial numbers on the ECU before your purchase.

Several modifications will need to be made to the engine harness in order for the vehicle to function properly with these computers. First off, the addition of the idle air control valve and cylinder position sensor will require additional wires to be added into the harness.

Other electrical modifications include the addition of the ECU memory, the ignitor signal, a vehicle speed

Often, distributor connectors need to be swapped out for those of the donor engine. Attach the plugs to the old harness by removing the pins or cutting and soldering the wires.

sensor (except on '86–'87 ECU), and the electrical load detector. Wiring modifications like these can be downright tricky. Unless you're very fluent with electrical schematics, it might be a good idea to contact a company such as Hasport Performance or Place Racing for one of their conversion harnesses. It just might save you from a big headache in the long run. It's important to note that all of this information is concerning the Si model Civic and CRX only. All other models will require even more wiring due to their carbureted fuel systems. So, take a word of advice and avoid these if you know what's good for you.

Anytime you're removing or installing an engine on these vehicles, it's always best to do so with the transmission side at a downward angle. You'll find that the engine won't fit if you try to remove or install it without an angle.

Installing the ZC

With the wiring taken care of, next on the agenda is to prepare the engine for installation. As with most engine installations, it's always a good idea to attach the wiring harness to the donor engine before it is placed into the vehicle. In addition, don't forget about the Integra engine mounts.

The use of the '86–'89 Integra or JDM Integra-style ZC mounts is mandatory in order for this engine to fit

Since the old Civic or CRX engine mounts are incompatible with the ZC engine, you might as well just throw them away along with the old engine and trans.

properly. With the exception of the driver-side engine mount, which can be reused from the original Civic, all of the other pieces will be from the Integra. Just make sure that the bracket that attaches to the driver-side mount is from the new engine. Were you to be working on a right-hand drive JDM Civic, then all of the mounts, including the driver-

side, would be Integra components. Although both the USDM and JDM chassis appear similar, there are slight differences underneath the hood concerning engine brackets.

Don't forget to install the new engine into the bay at an angle in order for it to fit properly below the frame rails. You can do this by using an adjustable load-positioning bar attached to the engine hoist. You'll quickly find that when trying to lower an engine and transmission down horizontally, that it just doesn't seem to fit from side to side. That is why always lower the transmission side down first.

Knuckles, Hubs, and Axles

Moving on to the lower portion of the vehicle, you'll soon see that there is quite a bit of work to be done with the axles and suspension. Although unusual for Honda and Acura engine swaps, the Civic front suspension will have to be swapped out in order to accommodate the larger Integra axles. Since either an Integra or an Integra-style ZC transmission is going to be used, you'll need Integra axles. They can be acquired from any '86–'89 Integra and can be installed without any modifications to the axles themselves. However, the suspension does need to be addressed. The front knuckles and hubs will both need to be exchanged for '86–'89 Integra units. Since the chassis of this particular Civic and the first-generation Integra are so similar, the process is a straight changeover.

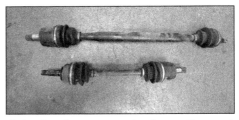

Selecting the proper axles is one the most important elements of the engine swap. The shorter Integra axle will replace the longer Civic one on the driver side.

Shifters and Cables Tech

With the installation of the '86–'89 Integra shift-linkage mechanism, the underside of the vehicle will be complete. You'll also need the clutch and throttle cables. The clutch cable may be found on any '86–'89 Integra with a manual transmission, and it goes in the same way your old one came out. You should purchase a new one, since they

Once the tranny is in place, you can reattach the Integra clutch cable to the bracket on the ZC transmission. You need to use this cable because the Civic cable is too small.

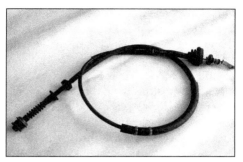

This right-hand drive throttle cable is way too long. It would need to wrap clear around the engine, and it won't sit properly. Be sure to install the proper USDM throttle cable.

do wear out and can eventually snap. You cannot simply reuse the Civic clutch cable, as it will soon slip off the bracket on the transmission due to the smaller plastic retaining nut.

You must also use the throttle cable from an '86–'89 USDM Integra. Right-hand drive throttle cables are usually way to long since they have to wrap around the engine.

If a vehicle speed sensor was added, don't forget to add the speed sensor cable from the Integra as well. Both the cable and the retaining clip will be necessary for a proper installation. Be careful not to lose the clip, as clips have a way of disappearing.

Fuel and Cooling Systems How-To

From cables to fluid systems, don't expect things to get any more complicated at this point. The fuel system is a

no-brainer, and you'll find that the injection feed line and fuel return line will both reattach in the same way that they were removed. In rare cases, the Civic may be equipped with an incompatible fuel injection feed hose due to differences in the hose end. In this situation, the best fix is to obtain the proper hose from the CRX Si model with the corresponding banjo hose end for the new fuel rail. Both hoses attach to the fuel filter in the same way.

The radiator, its brackets, and the cooling fan may all be reused since they don't interfere with the new engine. The original Civic heater hose may be reattached to the cylinder head; it is still plenty long enough to reach. Upper and lower radiator hoses may be created by trimming '86–'89 Integra components to fit.

A/C and Power Steering Made Easy

The Civic A/C compressor, bracket, and all of its lines can be retained. The power steering found on the rare GL model Civic can also be retained simply by reusing the Civic pump and attaching it to the ZC brackets.

Wrapping Up The ZC

It is very unlikely that any additional stopping power beyond the Integra brakes will be necessary for this vehicle. Since the ZC engine provides a modest 130 horsepower at most, you'll find that problems related to insufficient

braking will be rare. The same can be said for the handling of the vehicle. Since there has only been a 25-pound weight increase to the front of the chassis, adverse affects are extremely unlikely.

Remember that this car was originally equipped with this engine in Japan, so there aren't going to be any negative drivability issues if the correct parts are used. Since the engine is stock, if it is in good condition when installed, the owner can expect a totally reliable form of transportation with just a little extra kick. This is definitely the type of car that you wouldn't mind sending your mom off to the store in without worrying. It's rare that you'll find a reliable swap for a vehicle this old at a price that won't break the bank, but if there ever was one, the ZC swap is it.

B-Series Engine Swap

For those proud owners of any '84–'87 Civic, '84–'87 CRX, or '86–'89 Integra that demand more power than the ZC engine, you're in luck. You can thank the folks at Place Racing and Hasport Performance for making swaps into these older chassis possible. Their engine swap kits allow the B-series engine to be retrofitted into the third-generation Civic and first-generation Integra engine bays with relative ease as long as the procedure is being performed on a fuel-injected vehicle.

A skilled mechanic should complete such a project in just over a day. This is

Removing this fuel dampener pulsation cap with a 22-mm wrench will allow you to install the Civic fuel injection feed hose. Be careful not to over tighten this cap.

The ZC sits underneath the hood like it was always there, as a true bolt-in swap should. Although it looks stock, keep in mind that the JDM ZC will not be smog legal even though it's almost identical to the Integra D series.

Anytime you're doing a swap as custom as this, you'll need some aftermarket mounts such as these from Place Racing. OEM mounts cannot be used for these B-series transplants unless you want to do some cutting and welding.

similar to the timeframe you'd find in B-series engine transplants into other vehicles as well. Due to the variety of custom parts and slight frame modifications, it is important for an unskilled novice to research and make sure all of the proper parts are acquired before beginning the project.

B-Series Engines and Transmissions

Since these particular Civics and Integras do not use any type of late-model OBD electronics system, the B-series engine selection process is very important. For those folks wishing to go the non-VTEC route, the '90–'91 USDM Integra offers the original B18A1 drivetrain. All of the other non-VTEC B-series engines were produced after the implementation of OBD, with many featuring the hydraulic-style transmissions. Of course, your vehicle may be converted to OBD status, but with the availability of several excellent OBD 0 computers, there is really no need for a messy wiring conversion in most cases.

Other limitations to using the late-model engines include the fact that they're equipped with hydraulic-style transmissions, which aren't compatible with either of these two chassis. Only cable-style, B-series Integra transmissions will be compatible with these particular projects. These can be acquired from the '90–'93 USDM Integra.

Compatible VTEC engines include the B16A family, which can be found in

the '90–'91 JDM Integra RSi and XSi. Although this engine was in production for two more years, those later models were OBD I. A similar JDM, OBD 0 engine may be acquired from the '88–'91 Civic and CRX SiR, for example. The transmission that came with these engines will work equally well. If necessary, any of the late-model OBD I and II engines may be used, provided the electronics system is addressed and you use the cable-style Integra transmission.

With that being said, the B16B, Civic Type R engine, and the B18C GSR or Integra Type R engines may all be worthy considerations for these vehicles as well. As with most Honda engines, you get what you pay for. Of course, the older B16A and B18A1 engines mentioned first are the cheapest and easiest to find. When shopping for anything Type R you can plan on at least tripling your budget.

ECUs and Wiring for the B Series

If the vehicle is fuel injected and is to remain, OBD 0 leaves you with two possible electrical scenarios. In the first and less-complicated situation, you'll use an OBD 0 engine and computer. Compatible non-VTEC B-series ECUs can be found from the likes of the '90–'91 USDM Integra. These will be labeled as a PR4 unit. VTEC ECUs labeled as PR3 and PW0 may be acquired from the JDM Integra and Civic mentioned earlier if you plan on swapping a VTEC into place. You'll find that all of these computers will plug

into the vehicle harness with ease. You'll need to wire the cylinder position sensor, the idle air control valve, and the vehicle speed sensor into the Civic or Integra engine harness. Further modifications include provisions for the ECU memory, electrical load detector, and the ignitor signal. VTEC engines will need to add provisions for a VTEC pressure switch, VTEC solenoid, and a knock sensor.

The second scenario will involve the use of an OBD type engine and backdating its sensors in order to accommodate the OBD 0 vehicle and ECU. Apart from performing the wiring modifications to the harness just mentioned, several components on the new engine will need to be swapped out. The fuel injectors, distributor, and oxygen sensor will need to be replaced with units that can be acquired from an older OBD 0 B16A or B18A1 engine. With the addition of these components, you can enjoy the benefits of a newer engine without having to convert the entire electrical system of the vehicle over.

However, what if you want to use an ECU that is only offered in OBD form? What if you want to convert the car to OBD status? Well, first I'd recommend getting a new car, as the money you'll spend on an OBD engine could easily outweigh the value of the vehicle. Nevertheless, if there is a will there is a way, and it certainly can be done.

With the use of any OBD I or OBD II ECU, the proper connections must be made to the vehicle harness in order for the new computer to plug into place. Providing all the OBD sensors are present on the new engine, the resistor box may be removed from the chassis, meaning you need to rewire this portion for the new injectors. Injector plugs, distributor plugs, and oxygen sensor plugs are just a few of the connectors that must be swapped over in order for the new engine components to plug into place.

The fan switch must also be lengthened to reach over to its new location. For those using an OBD II ECU, be sure to select a JDM unit, as it does not require the additional wiring of all of the USDM OBD II emissions sensors. The same necessary VTEC modifications will need to be installed on all OBD

This B16B Civic Type R engine is one of only a few cases where an OBD II engine should be installed into these chassis. This scenario is all too rare though, as the engine costs twice as much as the CRX it's going into.

Almost all VTEC engines require a knock sensor. If you don't wire it up properly, you'll get a check-engine light and VTEC won't work correctly.

The distributor plugs shown here are just one example of several connectors that need to be swapped out for those of the B-series donor engine.

VTEC-equipped engines as with earlier VTEC ECUs.

Installing the B Series

Once the wiring is completed and installed onto the engine of choice, the selected engine mount kit may be attached to the engine and the vehicle. Engine mount kits are available for both the Civic and Integra to B-series engine swaps from Place Racing and Hasport Performance. These excellent mount kits place the engine in the most optimal position underneath the hood.

When using either of these products for the Civic or the Integra, you'll notice that they both include three mounts and either one or two adapter brackets, depending on the kit. Although very similar in appearance, the Civic and Integra kits differ slightly in the construction of the left-hand (driver-side) bracket.

This Hasport Performance left-side engine mount bolts perfectly into place. This is one of the only B-series engine transplants where the OEM bracket in the middle of the picture will not be used.

Once the mounts of choice are selected, the passenger-side bracket must be bolted to the right frame rail of the vehicle before engine installation. Next, the rear engine mount must be swapped out for the new unit and hand tightened into place. Last, the driver-side Hasport Performance or Place Racing mount must be bolted into place on the left frame rail. The procedure will be the same whether you're swapping into a Civic, CRX, or Integra.

Don't forget to make a large dent with a ball-peen hammer on the driver-side frame rail where the new alternator will rest. This is an important step for all of these vehicles to prevent the destruction of the alternator pulley. With the dent made, the engine must be bolted to the cable-style Integra transmission of choice. A '99–'00 Civic Si rear engine bracket can be finagled into place while lowering the

Alternator clearance is at a bare minimum. Installing a slightly longer belt will require you to adjust the alternator back closer to the frame, which is a bad idea.

There aren't any clearance issues around the engine bay. Notice the JDM Integra Type R tubular exhaust manifold tucked away.

engine in at an angle sloping down toward the transmission side. Once the engine is level, connect the rear engine bracket and tighten down the two side mounts to the specified torque rating.

A Lesson in Axles and Suspension

With the engine sitting in place and the engine hoist slid out of the way, it's time to get down and dirty and button up the underside of the vehicle. As far as the Integra goes, there isn't a whole lot more to do with the suspension other than to reattach it in the same way that it was removed. The Civic and CRX are an altogether different story.

Since the hubs on the Civic chassis are much smaller in diameter than those of the Integra, using the Integra axles isn't an option. If you don't want to make a set of custom axles to use the weak Civic outer joint, you'll need some '86–'89 Integra knuckles and hubs. You'll find that installing these components onto the Civic chassis is about as difficult as driving to the wrecking yard to pick them up. Thanks to the engineers at Honda, this will be a simple bolt-on affair.

Custom-length '86–'89 Integra axles may be purchased from Hasport Performance or Place Racing and will be required for both the Civic and the Integra. On the other hand, you could do the measurements yourself and have some axles custom made at any driveshaft rebuilding shop. Either way, you cannot use stock axles from any vehicle on this swap.

Custom-length axles will be mandatory for this swap, and the joint that attaches to the intermediate shaft may differ. Since this setup uses a late-model half shaft, it requires the corresponding late-model axle with the female end.

Shift Rods and Cables

As for the underside of the vehicle, a custom shift linkage must be purchased from the likes of Place Racing or Hasport Performance. Although a modified version of Integra and Civic linkages may be used together, this process involves some cutting and welding. With the custom shift linkage, simply attach your original shifter and shift knob and bolt the linkage into place. That's it. Upon test fitting, you might assume that the '86–'89 Integra linkage will work in conjunction with the B-series swap, but look closer and you'll soon realize that it's a bit too short. As with the Civic, a custom unit can be ordered, or the original piece can be cut, lengthened, and welded back together. You decide.

When hooking back up the various cables in the engine bay, you won't find too many surprises. The original clutch cable on the Integra may be reused, but the Civic will require a newer '88–'91 Civic cable. Failure to replace the Civic cable will result in possible cable slippage in the future since it isn't the right size. Both cables will attach up to the Integra-style transmission in the same way.

Since the '94–'01 Integra GSR intake manifold (which requires a different throttle cable) won't clear the vehicle's firewall, throttle cable selection will be kept to a minimum. The throttle cable for this swap is found on any '90-'93 Integra. This will work on all of the possible B-series swaps into these vehicles.

Integras and Civic Si's have the option of using a special throttle-cable

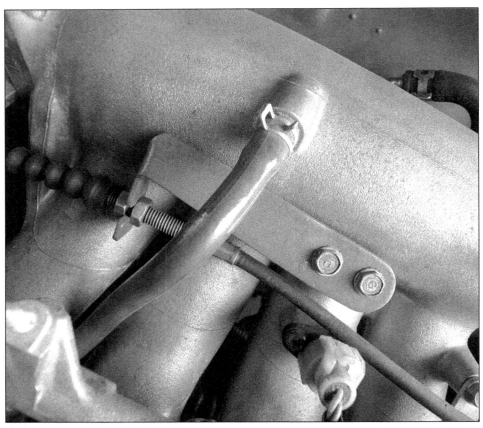

This special bracket from Place Racing allows you to reuse the original throttle cable on the Si models, which saves you from a bit of work underneath the dash.

bracket from Place Racing. You'll find that you won't have to crawl underneath the dash to remove the old cable. Simply attach the original cable to the new bracket and you're done.

Fluid Systems How-To

The fuel injection feed line situation doesn't vary much between the Civic, CRX, or Integra models. It's important

that the proper type of feed-line banjo is used, since all B-series engines use the later-style fuel-injection banjo fitting. Failure to install the proper fitting that attaches to the fuel rail will result in poor engine performance due to an inadequate fuel supply.

As is the case with most of the fuel-injected Civics and Integras, the original fuel injection feed hose will work

Custom shift linkage setups from Place Racing or Hasport Performance have OEM-looking ends that attach to the transmission.

Install the new Integra clutch cable underneath the dashboard of the Civic with this special clip. There's no way around this one; it must be swapped out.

The aluminum sealing washers on the fuel rail should always be replaced when they're removed. The smallest of fuel leaks can pose the biggest problems.

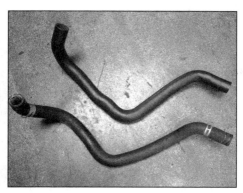

The newer Integra radiator hose featured on top is slightly larger in diameter compared to the Civic and Integra hose. The hose must be tightly clamped down to the radiator neck because of its larger size.

perfectly with the B-series fuel rail. In rare instances, certain Civics will be equipped with an incompatible fuel feed line. Be sure to replace this hose with a unit from the CRX Si. It's a good idea to always replace the original aluminum crush washers due to their inability to reseal.

Fuel return lines will reinstall in the same way that they were removed from the previous engine. In the case that certain B-series fuel pressure regulators provide a larger diameter outlet, modifications must be made to the original fuel return hose if its diameter is too small. A brass reducer/splicer will quickly remedy the situation. Consider the fuel system finished after you install the proper fuel-injection hoses and clamps.

As with any of the '84–'87 Civics and '86–'89 Integra to B-series conversions, any of the original radiators will fit, providing the proper hoses are used. You'll have to choose your upper radiator hose depending on which type of B-series engine you use. Non-VTEC models will require the use of a '90–'93 Integra RS, LS, or GS upper hose that is trimmed to fit. B16A, B17A1, and Type R engines need an upper radiator hose from any '94–'97 Del Sol DOHC VTEC or '99–'00 Civic Si. It also must be trimmed to fit. Integra B18C GSR engines will need to use a trimmed '94–'01 Integra GSR upper radiator hose. A lower radiator hose found on this same model GSR Integra will suffice for any B-series swap into these two chassis.

It's important to use adjustable hose clamps on both radiator connections in order to ensure a tight seal since the new hoses have a larger diameter than the radiator inlet and outlet. To finish off the cooling system, mount the original cooling fan just as it was removed.

Fuel Pump Upgrades

Anytime you're installing a more powerful engine into any vehicle, it is always wise to make sure that all of the existing components are up to the task. When dealing with a car that is well over ten years old, the fuel pump is certainly one area that should be addressed. Since many Civic fuel pumps were not designed to flow enough to support in excess of 100 horsepower, a replacement may be necessary. Many Integra fuel pumps are interchangeable with the Civic chassis and can be retrofitted by simply changing the electrical connector. Integra, Civic, and Accord chassis receiving VTEC engines may want to look to the aftermarket. Walbro offers fuel pumps flowing in excess of 250 liters per hour, which will quench any fuel shortage for all of these

transplants. Prelude owners may simply want to install a unit from the VTEC model, as these are adequate in most situations.

In many cases, changing fuel pumps will take no more time than an oil change. Most of the newer vehicles provide access to the fuel pump by means of simply removing the back seat and unscrewing a cover plate. In contrast, swapping fuel pumps on some of the older vehicles can be quite a chore, as the gas tank must be dropped in order to reach the pump. Although this can be quite time consuming without a vehicle lift, it's something that should definitely be considered once a more powerful engine is installed. It's a small price to pay to avoid the potential hazards of lean fuel mixtures.

This Civic fuel pump can be replaced with a high-performance unit by simply removing if from the housing that it is mounted to and slipping a new one into its place.

On the newer vehicles, you can simply remove this cover plate to expose the fuel pump housing. From there, just remove a few more nuts, pull the pump out, and you're done.

How to Keep Cool

As for keeping yourself cool, you might as well not get your hopes up. Due to incompatibilities with air compressors and engine brackets, not to mention a major lack of space, the ability to maintain the air conditioning system on these swaps is minimal. Although a B-series

A/C compressor can be finagled into place with some cutting, custom lines will be mandatory. Any '90–'01 Integra A/C compressor will attach to the block in much the same way. Custom lines can be made in order to reattach the fluid hoses to the new compressor. These will need to be professionally made due to the extreme pressures present in the system.

Unless you're prepared for quite a bit of cutting and fabrication, there is very little chance that a power steering pump is going to fit underneath this hood.

Stopping and Steering

Thanks to the four-wheel disc brakes found on the Integra, stopping issues will be of little concern post engine swap. However, the Civic is a different story. With the addition of at least 50 horsepower from the B series, stopping issues must be addressed. Luckily, the Integra knuckles you'll need to install also enable you to use the Integra brakes. Installing the larger Integra rotors and calipers will greatly reduce the stopping distance on the Civic.

Another concern for both of these cars is suspension issues. Due to the additional 100 pounds of new engine weight up front, understeer might occur. Aside from upgrading the shock absorbers and adding a stiffer rear anti-sway bar, there isn't a whole lot that can be done to remedy this problem. Until upgraded suspension parts are available, you just may have to live with this problem.

In Closing for the B Series

The B-series engine swap into this chassis isn't for everyone; it's definitely a transplant for the true straight-line enthusiast. Between the minor adverse steering effects and the elimination of air conditioning and power steering, this swap might scare away all but the true hardcore Honda freaks. Let's also not forget the fact that this engine is almost as big as the engine bay that it sits in. On a positive note though, with the addition of polyurethane engine mount kits, the ever present vibration inside the vehicle will always remind you of that V-8-killing beast that you have underneath the hood.

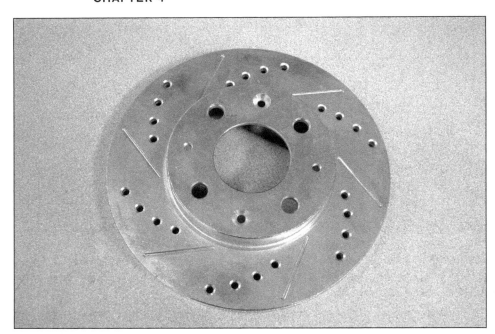

Cross-drilled and slotted rotors are an upgrade that you should look into when doing any engine swap. With all the extra power from the B series, it would be nice to know that you can stop it.

It's not too often that you'll see a Type R engine underneath the hood of the first-generation CRX. This car is sure to be a rocket considering its light weight.

1988 TO 1991 CIVIC

One of the first Hondas to receive an engine swap back in the early 1990s, the fourth-generation Civic has held on for the ride ever since. This particular generation of the USDM Civic officially arrived on the dealer lots with the '88 model lineup. The next body style change would not occur for another four years. Gas mileage and sales figures of the Honda Civic and CRX increased because they were produced only with programmable fuel injection. Quickly becoming one of the most popular cars ever sold, no one imagined how popular the ED/EE chassis Civic (commonly referred to as EF, but those are JDM only) would become among Honda performance enthusiasts.

Civic Offerings

With several different body styles to choose from, the '88–'91 Civic has a flair for everyone's tastes. The assorted Civic body styles built during this period include the three-door hatchback, the sedan, the wagon, and the sporty CRX. Interestingly, a coupe was not designed for the '88–'91 body style.

Several trim levels were available on many of these models, ranging from the top-of-the-line EX, Si, and 4WD wagon, all the way down to the stripped LX, DX, HF, and STD models, depending on the body style. The '88–'91 Civic chassis varies in its dimensions and weights with the lightest weighing in at barely over 1,800 pounds. In addition to its lightweight frame, the entire lineup sports a lower stance in comparison to its predecessors. Statistics like these make light-

Eric Cardines of Honda Fiend installs this early model B16A engine into a fourth-generation Civic hatchback with the help of a set of Place Racing engine mounts.

This well taken care of CRX is one of many examples of clean, pre-owned Hondas for sale. The lightweight HF model is a perfect swap candidate.

weights like the CRX HF such a popular transplant candidate. Remember, weight reduction is just like adding horsepower.

Another appealing characteristic of the fourth-generation Civic is its low price tag. Although well-kept Hondas will hold their value, engine swappable Civics with decent bodies and blown up engines may be purchased for next to nothing and are available all over the country. Other features that drew in more Civic-buying customers included optional power steering on the EX, LX, DX, and Wagon models, as well as air conditioning available on most all trim levels. Honda also introduced its development of the double wishbone suspension for the Civic, and a more potent disc braking system on the EX, Si, and 4WD Wagon models.

Although most Hondaphiles probably couldn't care less about what is currently under the hood of a stock Honda Civic, several engines made their way into this chassis. The EX, Si, and 4WD Wagon models share the same 1.6-liter engine otherwise known as the D16A6, while the STD models feature the D15A1 engine.

All other Civics built during this time period excluding the HF are equipped with the D15B2 engine. The CRX HF is the only one of the bunch that received the lowly D15B6. Of the entire bunch, the most powerful EX and Si models still only muster up a measly 108 horsepower at 6,000 rpm. It's safe to say that these cars are in need of an engine transplant.

OEM Upgrades

Many aftermarket parts are available for the Civic and '88–'91 CRX, but in most cases, you need only look to other Civics and Integras for high-performance upgrades. Take the braking system for example. One of the most common upgrades being performed on this Civic is the addition of rear disc brakes. You'll find that it is a true bolt-on process to swap the parts from a '94–'01 Integra. Just swap over the entire rear trailing-arm assembly, along with the complete brake setup, including the emergency brake cables.

Front brakes may be converted to high-performance status by means of assemblies found on the '90–'91 Civic EX. Using the knuckles, wheel hubs, brake lines, rotors, calipers, and pads from the EX makes this an easy, bolt-on swap. Any further braking performance will need to be sourced from aftermarket companies such as Brembo or AEM.

Suspension upgrades can be made to the '88–'91 Civic with assistance from the folks at Progress Suspension. Engineers at Progress have developed lowering springs specially designed for engine transplants that help counteract the common problem of understeer associated with the heavier engine. Sway bars, coil-over shocks, and lower tie bars are just a few other products that Progress has designed with engine swaps in mind.

OBD Options

When selecting the engine of choice for the '88–'91 Civic chassis, it's important to keep in mind that this vehicle is OBD 0. For the sake of simplicity when dealing with the electrical system, an engine manufactured before 1992 will make the most sense. If you're using an OBD I or OBD II engine, you'll have two options as far as the electrical system is concerned. First, it is always possible to retrofit the engine to OBD 0 status. You can maintain an OBD 0 ECU in the vehicle by backdating the sensors, fuel injectors, distributor, and emissions system of the newer engine. Going this route allows you to use a much newer, sometimes more-powerful engine, but doesn't require any major wiring conversions to be performed on the vehicle itself. In most cases, this is a perfectly acceptable alternative to a lengthy and labor-intensive wiring conversion. Your second option is to use the OBD I or II ECU.

Rear brake conversions on the '88–'91 Civic are a breeze, thanks to the Integra. The trailing arms are nearly identical so they'll simply bolt into place.

These dual-injected D-series engines are worthless as far as horsepower is concerned. The money spent on this header would be better used on an engine swap.

The Civic EX has much to offer to its lowlier counterparts. Its larger front and rear disc brakes are both compatible with all '88–'91 Civics for an easy conversion.

When Choosing OBD I or OBD II

If a compatible OBD 0 style computer isn't available for the swap of choice (such as the H22A or the SOHC VTEC) then converting the car to OBD is unavoidable. Depending on which OBD system is to be used, the appropriate ECU connector plugs must be retrofitted onto the Civic's underdash wiring harness. Look for the ECU underneath the carpeting on the passenger-side floorboard. Once the computer is unplugged, the original plugs must be cut off the vehicle harness. Remember not to touch the engine wiring harness at this time, as it must remain intact for its reuse on the new engine. The same process must be performed on a junkyard vehicle in order to obtain the new connectors.

OBD I plugs can be found on any '92–'95 Honda, while OBD II plugs can

Plugs like these aren't exactly easy to come by nowadays. Many wrecking yards are unwilling to cut apart a perfectly good vehicle harness for a few plugs. You might just have to buy the entire wiring harness.

be found on any '96–'01 Honda. Interestingly enough, even though they're both OBD II, the '99–'01 Civic and Integra ECU plugs are completely different from the '96–'98 versions. When converting to OBD II, it is wise to take the OBD II ECU of choice with you to the junkyard to make sure you get the proper connectors. Although these plugs cannot be purchased from the dealership or any auto parts stores, many wrecking yards will be more than willing to cut them out of the vehicle for a small fee. Until companies develop plug-in adapter harnesses like you'll find for the later-model Civics, this, unfortu-

Engine trouble codes may be viewed through the lower hole on the face of this early model ECU. This is how all OBD 0 computers transmit their codes.

nately, will be the only way to go. At any rate, once soldered into place, these new electrical connectors will allow the OBD style ECU to plug into place. It will be important to consult both the service manual for the Civic and that of the donor engine's ECU in order to make proper electrical connections.

Checking Codes

When converting an OBD 0 vehicle to OBD status, you're going to run into a small problem dealing with the ECU. In order to identify engine trouble codes on the earlier OBD 0 vehicle, Honda developed a flashing display on the front of the computer. Interpreting the number and length of the flashes indicated to the mechanic what the problem was. This display is no longer available if you replace your ECU with a newer OBD ECU. The OBD method for obtaining engine trouble codes is with a display on the gauge cluster via a signal sent from the ECU. You must wire this up appropriately if you wish to be able to diagnose future problems. Service manuals from both the '88–'91 Civic and that of the new computer will both be necessary to get this figured out.

Using the Original Harness

In most cases, you'll want to reuse the original Civic engine harness when swapping in a new engine. Right-hand drive JDM engine harnesses are worthless and need only be saved for a few necessary plugs and connectors. Most of the time, the engine wiring harness will need to be lengthened in various spots for the relocated sensors on the new

engine. Start by plugging the injectors into the harness first, thus giving a good starting point for which connectors must be lengthened. Take the necessary precautions when cutting connectors and lengthening wire to avoid mixing up the orientation of the wires in relation to the plug. If you don't, you'll need to consult the service manuals in order to pin them back into their correct locations.

ZC Engine Swap

The 1.6-liter DOHC ZC engine was one of the most sought after engines in the early days, but isn't the most common donor engine for '88–'91 Civics today. The introduction of VTEC didn't help this statistic much, but the fact that the ZCs that fit into this chassis haven't been in production for over a decade doesn't help either. It is also important to note that there is less aftermarket support for the ZC engine than any other in this book. The only qualities that allow the ZC to remain a viable swap through all these years are its versatility, ease of installation, and cheap price tag. The 2C swap can be accomplished by a novice mechanic in a day's work, with a minimum amount of difficulty and cash.

ZC Engines and Transmissions

In order for this swap to occur without a hitch, it's extremely important to select the proper ZC engine to begin with. Although very similar to the ZC,

This JDM ZC engine will make a perfect addition to the Civic chassis. In all cases, you can reuse the transmission from the original Civic on this transplant.

the D16A1 and D16A3 drivetrains available in the early model '86–'89 USDM Integras are in no way compatible with this chassis. For the correct ZC engine, you'll have to search out your local import engine supplier for the proper JDM version. The most popular choice may be found in select overseas '88–'91 Civic Si and CRX Si models.

European and JDM engines can be differentiated from the early U.S. version by their black valve covers (as opposed to some of the Integras having brown covers). Unfortunately, this simple color distinction alone won't ensure you of purchasing the proper ZC. In order to make sure you get the right one, it's important check where the driver-side engine mount is located. The correct ZC for this project will have its driver-side mount located behind the

Engine Perspective

"Install the new bracket on the front of the engine." Phrases like these can be confusing when dealing with the four-cylinder Honda and Acura engines. Although most know that the front of the engine is technically the portion of the block that houses the crankshaft pulley and the timing belt (or chain), many folks refer to the portion of the engine facing the front bumper as the front. In reality, the front of the engine is actually facing the side of the vehicle (except for the S2000). At times, this can make matters confusing when dealing with engine transplants. As you'll soon see, throughout the book the front and side of the engine are mentioned countless times. When I refer to the front of the engine, although not technically correct, assume that I'm referring to that part of the engine that faces the front of the vehicle. Sometimes that's just easier.

The front of the engine is actually the side that houses the camshaft gears, timing belt, and crankshaft pulley, all of which you can see here.

timing belt cover, similar to most late-model Hondas. If you find that this mount is positioned on the front of the engine, then you have the wrong one. This engine will work in an '84–'87 model Civic, but not in your '88–'91 EF chassis Civic. As with many JDM Honda engines, you'll find that most foreign wrecking yards usually have two or three ZCs in stock year round for very affordable prices.

When it comes to transmissions, unlike most other swaps, you're going to have a couple of choices here. Keep in mind that there are both Integra and Civic ZC engines, and that the transmissions are totally different from one another. The transmission of choice is the Civic-style ZC transmission available with the engine that is purchased. With the proper axles and intermediate shaft, this gearbox will bolt up and work perfectly. When going this route, it's mandatory to use the intermediate shaft found on the '88–'91 JDM Civic ZC; the USDM Integra version won't work.

The second option is to reuse the old transmission. That's right; this is one of the only Civic engine swaps where the old transmission will bolt up to a different engine. In fact, due to gearing preferences, many ZC swappers prefer the Si transmission to the ZC unit by far. You can save some serious cash by recycling the old transmission and axles. However, don't forget about the differences in clutches and flywheels. The clutch assemblies that you'll find on the

Depending on which year of Civic transmission you're using, these splines will differ slightly in their number and diameter. Make sure all your components match up when you're mixing and matching.

'88 engines aren't compatible with the '89–'91 transmissions (due to differences in the number of splines). Likewise, '89–'91 clutches won't work with '88 transmissions. Make sure that you're aware of the production year of the transmission you'll be using, whether it's a ZC or a Civic, before installing the clutch and flywheel. This doesn't mean that you must change the clutch in order to switch gearboxes, but it will be mandatory if they aren't compatible.

ECUs and Wiring for the ZC

In dealing with the wiring and electronics, the '88–'91 Civic to ZC swap requires a bit of work, but certainly isn't among the most difficult in this book. For vehicles other than the dual-point injected LX, DX, and STD models, the wiring remains reasonably simple. On all harnesses, because the cylinder position sensor is now located on the end of the exhaust camshaft as opposed to inside the distributor, the sending wires will need to be rerouted to the new location. In addition, the round distributor plug on the Civic engine harness (excluding LX, DX, and STD) will need to be cut off and exchanged for the square plug found on the ZC engine harness.

For those with the LX, DX, and STD models with the dual-point fuel injection, additional wiring will need to be performed in order to convert the dual-injector system to the regular four-injector setup. In addition to the two extra injectors being added, an injector resistor box must also be mounted and wired into place. Consult the service manual for schematics of the engine harness, and the wiring shouldn't take too

The cylinder position sensor runs off the exhaust camshaft on the ZC engine, as opposed to inside of the distributor on most others.

much time for the experienced. If you aren't interested in dealing with the electronics portion of the swap, Hasport Performance and Place Racing are just two of several companies that manufacture wiring kits to solve these problems. The addition of an EX, Si, or HF engine harness from the four-injector engines will also cut down some of the work for you if it's in the budget.

With all of the ZC engines being produced before the introduction of Honda's OBD electronics, finding a computer that will plug right in will be a snap. The installer has a few choices when selecting the proper ECU for the ZC swap. As in many cases, the most optimal ECU is usually the most difficult to find, and with this particular swap, things aren't any different. The computer of choice is a JDM-only unit, which can be found in the Civics and CRXs originally equipped with the ZC engine. This computer can easily be distinguished by the serial numbers bearing the code PM7 in the center. An almost equal alternative is the PG7 computer box available in '86–'89 Integras. The last option for a compatible computer can be found almost anywhere for dirt-cheap. The '88–'91 USDM Civic PM6 computer from the multi-point (four-injector) models is a viable alternative for those on a strict budget or as a temporary standby for the ZC unit. This ECU will work fairly well, but keep in mind that the fuel maps in the ZC computers will far better suit this more powerful engine.

Installing the ZC

Once the harness is installed and the transmission is bolted into place, it's time to select the proper engine mounts. Speaking of engine mounts; it really doesn't get much easier than this. All of the Civic brackets and mounts will work with the ZC engine block and transmission. This includes everything from the side, front, and rear mounts. With these fastened properly in place, the engine can be installed much in the same manner as the Civic engine was removed.

This transplant requires no cutting or modifying of the chassis in any way. With the installation of the proper

This rear engine bracket taken off of the Civic transmission can be reused regardless of the transmission type. The original Civic and ZC transmissions will both bolt up to the OEM rear mount by using this bracket.

The ZC engine will sit perfectly underneath the hood if you use the appropriate OEM Civic mount. Both the engine mount and the pocket mount remain unaltered when installing the ZC.

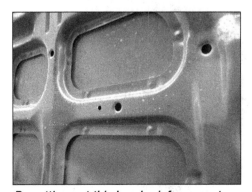

By cutting out this hood reinforcement webbing with a small hand grinder, you will allow enough clearance so that the ZC valve cover will not rub.

A perfect fit. The ZC engine sits underneath the hood of this Civic in the exact same spot as the old D series.

brackets and mounts, you'll notice that there is still plenty of ground clearance because the size of the engine block and oil pan hasn't changed much. Under the hood, things are just a little bit tighter than before. The first thing you'll notice is how close the timing belt cover is to the underside of the hood; it will indeed rub with a bit of engine vibration. Solutions to this problem include removing the upper timing belt cover, or spacing the left side mount down with a couple of hardened washers.

Axle and Suspension Options

Finishing up underneath, the reinstallation of the suspension and the axles isn't going to differ any from the removal of the stock units. In fact, if the Civic transmission is being reused, then both the original axles and all of the suspension components may be reinstalled along with it. If the ZC transmission is being used, however, then custom axles will be mandatory. When going this route, HF and STD models with the smaller wheel hubs will need to swap over to the larger units found on other models.

Along with the custom axles, you'll need a JDM ZC intermediate shaft due to the shorter Integra-style axle now being used on the driver side. Axles may be custom made by obtaining two of the original Civic passenger-side axles and modifying them by installing both of the '90–'93 Integra inner joints in place; one on each axle. Before installing the axles, the dust rings will need to be removed from the inside of each of the two wheel hubs. Failure to follow this step will pre-

You must remove this ring from each of the wheel hubs before installing the axles. If you forget, it will bind up against the axle, prohibiting the wheels from spinning freely.

The original throttle cable found on the ZC engine works fine providing that you use the Civic cable bracket. This is the easiest way to go.

vent the wheels from spinning freely. Since you'd need to obtain an extra Civic axle and a pair of Integra axles to make the homemade units, this is sometimes more costly than having a set custom made from a professional driveshaft builder.

Shifters and Pedal Controls

Before the car can be set down on the ground, the shift linkage must be installed. No matter which cable-operated D-series transmission is being used, the original shift linkage can easily be installed right back into place without any modifications. The original clutch cable can now be reattached back to the transmission-housing bracket and adjusted properly. To finish off the pedal assembly, the addition of an '88–'91 Civic Si throttle cable will ensure proper throttle control. This step, of course, is only necessary for those with the non-Si models.

Fuel and Cooling Systems

Moving on to the fluid systems, the fuel injection feed hose and return line can be reconnected using the factory lines. In cases where the vehicle is equipped with the wrong fuel injection feed line, it may be swapped out for that of the CRX Si. The feed line should use the large damper pulsation nut as opposed to the smaller cap nut when being fastened to the rail.

You'll find that the cooling system isn't much more difficult. Using the original Civic radiator and cooling fan, the original, unmodified lower radiator hose will slip on ever so easily. As far as the top radiator hose is concerned, one from the '86–'89 Integra will fit just fine once it is trimmed to size.

Filling the Trans

Filling the transmission fluid by means of the factory oil fill hole requires the use of a hand pump to feed the fluid in. When on your back, this can be a messy process to say the least. Rather than following standard protocol according to the Honda service manual, there is another way to simplify the filling process. Simply remove the vehicle speed sensor (VSS) on top of the transmission, attach a hose to the end of a funnel, and stick the other end down to the VSS location. Pouring the required fluid in from up there sure beats crawling underneath the vehicle.

When reinstalling the VSS, be sure that it sits flat before tightening down the bolt. If it doesn't, then the bracket shown on the right of the unit can snap off.

Of the three bolts that you can see in this photo, only the one off to the far left needs to be removed. Removing the other two bolts won't remove the VSS, but rather will take it apart.

Reinstall the radiator into its original location along with its brackets and cooling fan. It's best to remove it during the swap to avoid damage.

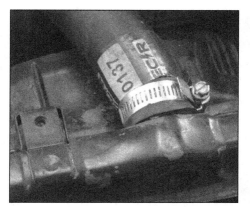

Since the ZC upper radiator hose is the same diameter as the Civic radiator outlet, you can use either an OEM or adjustable hose clamp.

Although not emissions legal in many states, the ZC engine swap is a popular alternative for enthusiasts on a budget. Notice how much clearance the ZC leaves on all sides of the engine bay.

A/C and Power Steering Tech

Air conditioning can easily be retained on the ED/EE chassis with a ZC transplant with a little effort. Since the ZC block shares many characteristics with the Civic block, the ZC is more than willing to accept the Civic compressor and bracket. Simply bolt up the bracket to the block, and then attach the compressor.

Even the power steering can remain intact, that is, if the vehicle is one of the rare few to be equipped with it in the first place. If so desired, the Integra or ZC power steering brackets can be used with the original Civic power steering pump. Without any tricks or modifications, the original pump will bolt right onto the brackets and work just as it did previously. Finish it up by connecting the original feed line.

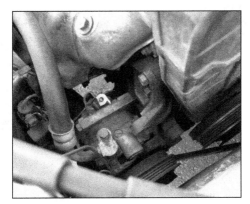

Air conditioning is a bolt-on affair providing that you have all of the proper components. The belt has yet to be installed on this transplant.

Post-Swap Options

Due to the lack of aftermarket support, you'll find that the most modified of ZC engines are generally pushing less than 175 horsepower. With this in mind, upgrading the braking system significantly is usually unnecessary. As long as the original brake components are in good working order, you'll find that they'll be more than adequate. Of course, other upgraded brake components can change over, but they're certainly not mandatory. In addition to stopping the vehicle, sometimes a significant change in the weight over the front axles on a front-wheel drive car will cause the vehicle to steer improperly. Situations like these are referred to as oversteer or understeer. Since the ZC engine weighs only a measly 25 pounds more than the Civic engine, understeer problems relating to extra weight should be non-existent.

ZC Wrap Up

You'll find that for the not-so-diehard Honda enthusiast, the ZC engine swap is a very acceptable alternative to some of the more expensive, larger-displacement engines. For a little extra kick in the pants without sacrificing any of the reliability you've come to love, these engines will still work great, even after years of neglect.

D-Series Engine Swap

You might be wondering why anyone would ever go through all the trouble of installing a measly SOHC VTEC engine into the '88–'91 Civic. I mean, why wouldn't you just do a B series? Believe me, after all of the customers that I've had inquiring about this transplant; I've wondered the same thing many times. The fact is, though, that this is a swap worth mentioning, and there are definitely enthusiasts out there who are doing it. It's easy to lose sight of how well engineered these single-cam engines are with the H22As and the B18Cs hogging so much of the limelight, but the fact is that these are great low-cost alternatives. Even better, these are advanced enough to be equipped with the VTEC system.

For most folks, the first attraction to the idea of transplanting a D-series VTEC engine into the fourth-generation Civic is the complete ease of installation. From a mechanical standpoint, this thing will virtually bolt right in, almost as easily as the other one is pulled out, making it an easy swap to do in your own garage. This is true only to a certain point. From an installation standpoint, the difficulty level is extremely low and it isn't uncommon to see a novice swapper perform this portion of the swap in a day's work.

SOHC VTEC engines are easy to find, yet they aren't available out of that many different vehicles. This 125-horsepower example came from a '93 Civic Si.

The original cable-style Civic transmission is perfectly compatible with the newer engine. Reinstalling it is a simple bolt-in procedure.

The new engine has its MAP sensor on top of the throttle body, so lengthen the appropriate wires from the vehicle harness to reach.

However, just because the installation is easy, unfortunately, doesn't mean that this swap is as simple as it sounds. In fact, one portion of the swap can be downright difficult. We'll talk about this more in a bit.

D-Series Engines and Transmissions

When considering which SOHC VTEC engine to swap into this Civic chassis, it really comes down to only two choices. The first engine candidate is that of the '92–'95 Civic Si hatchback, the Civic EX coupe and sedan, and the Del Sol Si. This engine is more commonly referred to as the D16Z6, and is typically found only in wrecking yards that carry USDM engines.

The D16Y8 from the '96–'00 Civic EX coupe is the second choice. Both of these engines are virtually identical, although the D16Y8 is OBD II. It is equipped with a different ECU, distributor, and injectors, among other components. You could also use a JDM SOHC VTEC D15B engine available from the likes of various Civics and CRX del sols, but they're usually rare here in the United States. With the D series, you have the choice of running either of the two OBD styles of engines, and it really won't make any difference mechanically. The OBD I computer is your best option from an installation standpoint. Fortunately, either engine and most of the electrical components will be totally compatible with the OBD I computer. These USDM engines are usually very easy to find and since this transplant does not require a transmission swap,

the cost can be kept to a bare minimum.

With the engine picked out, transmission selection will be next on the agenda. The number-one recommendation for this particular transplant would be the '88–'91 Civic Si or CRX Si gearbox. Since this tranny has an optimal gear ratio for racing and a direct bolt on fit, it is a much better alternative than retrofitting the hydraulic D16Z6 or D16Y8 transmissions into place. Although that can be performed with much fabrication, it is certainly a more time consuming and more costly route that just isn't necessary.

Once you've found a transmission, determining which clutch and flywheel you need will be next. Since the '88 Civics use a smaller, weaker clutch and flywheel combination, the '89 and newer Civic-style clutches are preferred. Since the number of splines on the clutch discs is different between the two styles, it's mandatory that the transmission matches the clutch disc. Long story short, don't expect an '89 or newer clutch to work with an '88 transmission. It's better to find this out now by test-fitting the components than after the swap is done and your transmission won't shift.

ECUs and Wiring For the D Series

With a cheap initial cost and a direct fitment up to the old gearbox and the chassis, it's no wonder that these swaps are as popular as they are. But, wait a minute. Unfortunately, it isn't just that simple. You now have a small problem with the electrical system. In most cases, when you install OBD I or

OBD II engines into an OBD 0 vehicle, you'll simply use the proper OBD 0 ECU that is compatible with the engine and perform the necessary wiring modifications. The problem is that there is no SOHC VTEC OBD 0 computer that works for this transplant. Honda did not develop this engine until '92, and by that time OBD I had already been fully implemented.

At this point, you're looking at two solutions to the problem. The easiest would be to reuse the original Civic ECU and install an RPM-activated switch that will activate the VTEC mechanism automatically every time the engine hits the desired RPM. Although many find this to be a viable solution, this remedy isn't going to be acceptable here for one major reason. On a Honda computer that is equipped with the VTEC mechanism, the ECU adds additional fuel through the injectors when VTEC is engaged. When using the RPM switch, VTEC will turn on and sound like it's functioning, but without the additional fuel, the extra power will be missing. Since losing power isn't an option, this leaves only one solution. The vehicle must be retrofitted to OBD I status.

We won't discuss converting the vehicle to OBD II status due to the emissions sensors that must be added to the vehicle, which in turn results in much unneeded labor. Normally, seeking out a JDM OBD II ECU that does not use these extra sensors would be smart, but since one doesn't exist, we'll skip OBD II altogether. Folks may still use OBD II engines, so long as an OBD

All SOHC VTEC engines are OBD I or II. That means you'll need to modify the late-model distributor for use with the old engine harness. Simply swap the plugs from the donor engine onto the Civic harness.

I computer is used with them. I told you that this swap wasn't all that easy; the wiring can get rather complex.

Converting a car such as the '88–'91 Civic (ED/EE) to OBD status will first require modifying the electrical plugs on the end of the vehicle wiring harness that plug into the ECU. Assuming you'll be using an OBD I ECU, these must be swapped out with plugs from any OBD I Honda or Acura. This is a very complicated and time-consuming procedure and isn't recommended for the novice auto electrician. After this procedure, the ECU will be able to plug into the underdash harness. Underneath the hood, a few minor modifications will need to be performed to complete the OBD conversion.

Just as with any other swap, the original Civic engine harness is to be reused and should be attached to the engine at this time. The injector plugs,

Since the fan switch is in a different location on the donor engine, you'll need to lengthen the wires from their original position, which was on the back of the engine block.

distributor plugs, reverse light plug, and oxygen sensor plug will all need to be cut off and replaced with the late-model connectors found off of any corresponding OBD harness.

You'll need to lengthen the plug for the fan switch and swap its connector for that of the donor. The fuel-injector resistor box must also be removed, as it is no longer necessary. Again, consult the service manual for wiring schematics and pin locations.

Excluding the EX, Si, and HF Civics, two additional fuel-injector provisions must be added into the engine wiring harness on all of the dual-injected engines. An EX, Si, or HF engine harness can be used as a source to help convert dual-injected engines. Finally, VTEC must be wired up accordingly in order for it to be activated at the proper RPM. The addition of wires for the VTEC pressure switch and the VTEC solenoid will both be necessary. With these wiring modifications completed, the P28 computer found in the '92–'95 Civic EX and Si models will work flawlessly. Since these computers are also being converted to work on DOHC VTEC engines, they're rather hard to come by nowadays, but they're out there. If wiring isn't your idea of a good time, custom harnesses may be purchased from Place Racing. Sometimes being able to plug right in and go is the best route if you're in a rush or simply don't want to screw things up.

Installing the D Series

With the harness plugged into the donor engine, it's time to install the engine. Before installing the new drivetrain, a couple of key modifications need to happen. First, you need to identify and install the proper motor mounts. Since the original transmission is to be reused, the old rear engine mount, rear bracket, and right-side engine mount can all be reused. As for the other side, the left bracket must be removed and replaced with the bracket off the original engine. Replacing this bracket allows you to bolt the engine up into the original rubber mount on the driver-side frame rail.

It will fit if you trim a little off the plastic timing belt cover. Once the engine is installed, the original front rubber

A combination of the '92–'95 Civic engine bracket and '88–'91 Civic engine mount will adapt this late-model D series into the older chassis.

If you reuse the original Civic transmission, you can slide the original front crossmember and front motor mount back into place.

mount should be reattached. In most cases, this mount will be broken, giving you a perfect reason to break down and buy a new one.

Now that the engine is sitting in place, it's pretty clear how well the D16Z6s and D16Y8s fit. With plenty of clearance under the hood, as well as the same amount of room in the engine bay as before, the addition of a turbocharger or a supercharger won't be hindered by a lack of space. Since the original mounts are used, ground clearance will be kept to a maximum as well, allowing for ample clearance between the oil pan and the pavement. As you'd expect, frame clearances will also remain the same. All of this clearance without the need for cutting or welding of any form, contribute to making this process a simple one.

Axles, Shifters, and Cables Simplified

Finishing things up underneath is going to require installing the proper

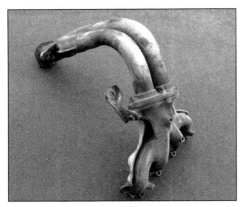

Although the original Civic exhaust manifold will bolt into place, this late-model unit from the D16Z6 VTEC engine will provide the utmost in OEM performance.

axles, suspension components, and shift-linkage mechanism. On this particular transplant, all of the original parts may be retained without modification. Simply reinstall the axles the way they were removed, which isn't too difficult. The original shift rods may be reused on all manual-transmission models. The original clutch cable may be reused as well, but in order for the gas pedal to function, you'll need a throttle cable from a '92–'95 Civic EX or Si.

Injection and Cooling Practices

The fuel feed line may be reattached provided the fuel injection feed hose is compatible with the new fuel rail. Depending on the donor engine, the fuel rail may be equipped with either the large damper pulsation nut or the smaller cap nut. In order to use the existing fuel line from the Civic, you need to

Swap the proper fuel injection feed line into place for the fuel system to work properly. The old line was incompatible with the damper pulsation nut shown here.

be sure that the same type of sealing nut is used on the donor engine, since it will be used again. Unfortunately, you cannot simply swap nuts to connect your fuel feed line; instead, you'll have to change fuel rails or fuel feed lines.

You can reinstall the radiator, cooling fans, heater hose, and lower radiator hose, as they're all still a perfect fit. Add an upper radiator hose from a '92–'95 Civic EX or Si to complete the cooling system.

A/C and Power Steering Solutions

If you're wondering, it isn't too difficult to retain the vehicle's A/C system. Since both engines are D series, the brackets and compressors will easily change over. The only tricky part is knowing which one to use on this particular application. The original A/C compressor may be used with the D16Z6 or D16Y8 A/C bracket. Once set into position, you'll see that the hoses and lines fit perfectly. The only problem that you'll find is a slight clearance issue between the oxygen sensor and the condenser fan. Minor trimming on the condenser fan will provide enough clearance.

The retention of the power-steering system isn't as easy, though. There's a conflict between the available pumps and brackets for these later-model engines, and there's only one way to solve it. You need a '92–'95 Civic power-steering bracket, but the power steering pump needs to be used as well. Once bolted into place, you'll find that the late-model pump isn't compatible with the '88–'91 chassis power steering feed line. The easiest solution involves some minor fabrication. The original line must be cut off about 4 inches from its pump end. Perform this same operation to a hose from a '92–'95 model Civic as well. Weld the new hose end to the end of the original power steering line. Since power-steering lines accommodate extreme pressures, it's imperative that the line be TIG welded by an expert. That's it; you now have a custom solution to your power-steering problem. Of course, if you don't have access to a welder, a shop specializing in high-pressure hoses can always make a custom line.

D-Series Satisfaction

Luckily for those interested in going the SOHC VTEC route, there really aren't any negative side effects. Since the two engines weigh roughly the same, braking and handling won't be affected.

If you do the wiring right, this is probably one of the most reliable engine swaps in this book. There aren't any axle problems, vibrating motor-mount issues, or ground-clearance dilemmas. The only thing that you'll notice from this swap is the additional power and smooth drivability of a VTEC engine.

Although not considered a true engine swap by some transplant enthusiasts, the bottom line is that these 125-horsepower ZC alternatives will more than likely satisfy the average budget-minded EF Civic owner's need for speed.

Eric Cardines of Honda Fiend is seen here installing a set of Integra rear brakes and trailing arms. Although it isn't necessary on the D-series swap, a brake upgrade is never a bad idea.

The appearance hasn't changed much since the D series was taken out a few hours ago. You'll be hard pressed to know that an engine swap has taken place unless you really know what you're looking for.

B-Series Engine Swap

No discussion of engine swaps would be complete without dealing with the B-series engine transplant into the '88–'91 Honda Civic chassis. Originally equipped from Honda with a B16A engine and transmission, top of the line early model Civics can be found virtually everywhere but in the United States. With the chassis begging for a B series drivetrain, it's no wonder that these engines are among the most popular swaps. This B-series swap is in the middle range of difficulty, and can be completed in a long day by a professional or as little as a couple of days by a novice. Although the wiring can get a little tricky on some of the base models, with the proper guidance, the mechanical portion of this transplant is a breeze.

B-Series Engines and Transmissions

More than any other engine that you'll read about in this book, the possibilities for B-series swaps for this chassis can be overwhelming at first. A few important JDM B16A engines are equipped with the "cable tranny" with which this type of Civic was originally equipped. The first is found in the '90–'93 JDM Integra XSi and RSi; the '90–'91s are OBD 0 and the '92–'93s are OBD I. Unless a time-consuming OBD conversion is desired, it's best to stick with the early models since they're all virtually the same engine.

The second compatible B16A engine is found in the '88–'91 JDM Civic and

Visiting the Engine Yard

With so many engines and transmissions, it can be downright overwhelming to visit the wrecking yard in search of the perfect donor engine. Upon visiting an engine yard such as Go Motors in Anaheim, California, you might be surprised at the never-ending sea of DOHC VTEC engines. It can be difficult to distinguish one from another with so many available. Assuming you have already narrowed down your engine selection to a model and year, there are a few other criteria that you should be aware of before your purchase. Following these simple guidelines just might save you a few bucks in the end.

By looking at the underside of the oil cap, often you can tell the history of the engine simply by the color and texture of the oil. Oil that looks like caramel is a sure sign of an oil and water mixture. Cases like these are usually representative of a failed head gasket or worse. The oil cap can also give you hints as to how the motor was cared for. Black sooty oil is usually a good indication of an oil changing neglect on the part of the previous owner.

When dealing with a reputable engine yard, most engines will be compression tested before being sold in order to ensure decent internals. As well as looking internally, you will not want to overlook the obvious. Be sure to inspect the entire engine block and cylinder head for cracks or any other type of damage. Often, wrecked vehicles provide a slightly cracked engine block unbeknownst to the junkyard employees.

Quality yards such as Go Motors thoroughly inspect their engines before their sale. Apart from a lot of dirt and grease, don't let the battery acid stains scare you into thinking that you won't be getting a decent engine. Frequently, some of the cleanest engines internally are some of the dirtiest externally. To affirm that your prospective engine is clean inside, be sure that the junkyard hasn't left anything open on the engine. Most of the time they will plug all of the holes closed with a cap or plug of some sort.

Once you've found an engine that meets your criteria for cleanliness, you'll want to make sure that it is equipped with everything that it should be. Most engines are sold with all of their mounting brackets and accessory brackets. Be sure that these are on the engine before taking it home, as you'll eventually need to have some of them. In addition to the brackets, other necessary components include a downpipe and exhaust manifold. Many engine yards do not sell their engines with this equipment, so be sure to ask for it. Upon the purchase of your engine, double-check all of the accessories and manifolds for tightness. Often, parts are swapped onto an engine at the wrecking yard in order to make it a complete package. It's best to make sure that these components are fastened properly now instead of once the car fails to start as the result of an intake manifold leak. Other concerns include the throttle position sensor, or the lack there of. Over half of these are usually broken on junkyard engines, so be sure the engine you choose has a decent one.

Having been in the engine business for over 16 years, James Go of Go Motors knows Honda engines. With hundreds of engines in stock, Go usually has just what Hondaphiles are looking for.

Many engines arrive from Japan in the form of a front half of a car, otherwise known as a clip. These are cut in two and shipped throughout the United States.

Notice the tape that covers up the fuel return line, the vacuum ports, and the throttle body. You might want to think twice about purchasing an engine with openings that expose the internals to the elements.

This JDM B16A engine is as complete as they come. Notice how it is equipped with a complete exhaust manifold, both radiator hoses, and of course, the transmission. The distributor has been removed so it doesn't get damaged.

You can identify cable-style transmissions by the bolt pattern for the top mount. The cable-operated arm in the upper portion of the photo also gives it away.

CRX SiRs. These engines are all OBD 0 and are extremely common in import wrecking yards across the United States. Similar in many ways to the early B16As is the B17A1, found only in the '92–'93 USDM Integra GSR. Although it's an OBD I engine, it's equipped with the cable-style transmission and can be retrofitted into any EF Civic, providing you're up for an OBD wiring conversion.

Other B16A donor engines can be found in the Del Sol DOHC VTEC and the '99–'00 USDM Civic Si, just to name a couple. Both of these B16As (including any of the ones you'll read about in later Civic sections), all use the hydraulic-clutch-style transmission. These motors are all equipped with OBD as well, some having a bit more horsepower, and are generally much more expensive and harder to locate than the older engines. Usually, unless you're installing a rare B16B Civic Type R engine with 185 horsepower, it's best to stick with the early model transplants, which are much easier from a wiring standpoint.

Often referred to as the big brother of the B16A, the B18C is the next step up for torque and horsepower in a B series. Since B18Cs don't come with anything but the hydraulic-style transmission, a conversion kit from Place Racing or Hasport Performance or using an older-style transmission are the only two possible options.

Fortunately, for those who do not wish to use the newer-style hydraulic transmission, the older cable-operated units bolt perfectly into place on any B-series engine block. In most cases, this is actually the preferred way to go since most of the B-series motor-mount manufacturers design their kits for the old cable-style units.

B18Cs are found in many '94–'01 USDM and JDM Integras. The more-powerful, race-inspired Integra Type R engine can be located in Japan on select '94 and up models, as well as in the United States on certain '97 and up models. Since all of these engines are only available in the OBD form, an OBD wiring conversion will be necessary in order for the engine to function properly with the OBD ECU. Although the market changes from year to year, for the most part all of the B18Cs are in very high demand and among the most expensive of all Honda and Acura engines; especially the Type Rs.

Losing popularity more and more every year, the non-VTEC Integra and CRV engines struggle to be serious candidates as donor engines. The one attribute that has kept these engines in the game for as long as they have been is the possibility of being converted to VTEC status in the future; otherwise known as an LS/VTEC. Many enthusiasts prefer the inexpensive price and abundance of the inferior non-VTEC B18 engine blocks, hoping to retrofit the VTEC cylinder head into place for a low-budget B18C alternative. Applying this same technique to the B20B CRV or B20Z S-MX engine blocks with a standard 2.0 liters of displacement will produce torque results that one would never find on any stock B18C. These particular engines can be found on the '97–'00 USDM CRVs and the '99–'00 JDM S-MXs, respectively.

The B18 VTEC alternatives include the B18A1 and the B18B1, with the latter being found on '94–'01 USDM Integras and select '94–'95 JDM ESi models. The B18A1 was introduced in 1990 and was present in the RS, LS, and GS models until the demise of the second-generation Integra in 1993. The first two years are OBD 0 engines while the last two are OBD I. All four years of the B18A1s use the cable-style transmission. As mentioned, the B18B1s were introduced on the new '94 Integra in the RS and LS models, and remained until 2001. All B18B1s are equipped with the hydraulic-style transmission and switched to OBD II in 1996. As in any other case, the cable-operated transmissions will prove themselves to be the easiest to install, as will the OBD 0 engine and computer. In the case of the later model B18B engines, the Integra OBD 0 ECUs have succeeded in working flawlessly in all cases, providing you backdate the sensors.

ECUs and Wiring for the B Series

You have several choices when shopping for a computer to go with that new engine. For any of the non-VTEC engines, the PR4 ECU found on all of the '90–'91 Integras will work perfectly. For OBD 0 VTEC engines, both the PR3 and PW0 ECUs found on the JDM B16A engines will be your best bet. When dealing with all other OBD B-series engines, it's usually best to stick with the ECU that belongs with the engine of choice. If you plan on reprogramming your computer in the future though, you'll find that it really doesn't matter a whole lot which DOHC VTEC ECU you use. If a GSR engine is being used and the secondary intake butterflies and the evaporative purge-control solenoid are to remain intact, remember that only the P72, '94–'01 Integra GSR ECU offers this option. Keep in mind that, wiring up the secondary intake system is optional, and leaving it un-wired won't pose any problems.

The intake air bypass valve found on the later-model GSR engines is an optional hook up. Although it isn't difficult to wire into place, the IAB won't pose a problem if you leave it unconnected.

When modifying the wiring harnesses for an OBD 0 B16A engine swap, the level of difficulty varies a little bit depending on the engine and vehicle combination. Non EX, Si, or HF '88–'91 Civics featuring the dual-point fuel-injection will need to first be converted to multi-point fuel injection. In contrast, those lucky enough to own the higher-end models won't need to worry about this, as these cars are already wired up for four fuel injectors. Prior to doing these modifications, dual-point injection owners may choose to purchase an EX, Si, or HF engine wiring harness from the junkyard, which greatly simplifies the rest of the process. If this isn't in the budget, then those additional injector wires will need to be added to the dual-point harness at this time. An injector resistor box found on

If your engine doesn't already have one, you'll need to mount an injector resistor box in the factory location up in the driver-side corner of the engine bay. Some dual-point engines don't come with them.

the multi-point injected models will also need to be mounted to the chassis and wired up to the harness. This resistor box is necessary in order for the injectors to function properly with the ECU.

On dual-point injection models, the crank angle sensor in the new distributor must be wired up at this time as well. All Civic models at this point will also require the addition of a wire for the second oxygen sensor in order for it to be recognized by the ECU. For all B-series engines, the Civic harness must be placed on the new engine and stretched or lengthened as necessary for it to fit properly. If a VTEC engine isn't being used, the wiring is complete at this time. For VTEC engines, the additional wiring of the VTEC pressure switch, VTEC solenoid, and the knock sensor will all be mandatory. OBD conversion swaps will require these same operations to be done to the engine harness as well as the conversion to the vehicle harness and additional OBD engine harness modifications mentioned in the '88–'91 Civic section. For those not wishing to partake in any of the wiring, Hasport Performance and Place Racing both manufacture plug-in wiring harnesses suitable for most B-series transplants.

Installing the B Series

With the wiring harness completed and connected to the B-series engine of choice, the engine mounts may now be set into place and the engine prepared for installation. Depending on which type of B-series engine you choose, the driver-side engine bracket may need to be replaced with a '90–'93 Integra engine bracket. This post bracket found on later-model B-series engines is designed for the newer-style chassis and won't work on the ED/EE Civic. The necessary rear engine bracket can be acquired from the same year Integra as mentioned above. Even with these two pieces, a B-series engine mounting kit from the likes of Hasport Performance, HCP Engineering, or Place Racing will also be necessary. With all of these kits, the replacement of the Civic's rear engine mount is mandatory.

When using the HCP Engineering rear mount, the lip on the crossmember

This HCP Engineering rear engine mount has been bolted into place after the rear crossmember lip was pounded down flat. It's better to flatten it too much than not enough so that the bracket won't fit.

must be folded down with a hammer in order to provide the appropriate clearance for the rear engine bracket. All mount kits require you to modify the frame for the new B-series alternator. The frame must be indented slightly with a large ball-peen hammer in the area where the alternator is to be placed in order to allow clearance for its pulley and belt. Once installed, the shortest alternator belt available must be used in order to avoid frame contact. After you've completed these modifications, the engine may be hoisted down into position.

Once the engine is lowered into place with an engine hoist, the aftermarket left-hand mount of choice must be attached to the post mount on the engine and set into the bracket on the frame rail. With the left mount attached, the rear engine bracket may be bolted

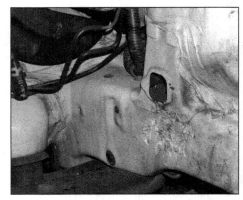

Right in the center of the photo, you'll notice where the frame was pounded for alternator clearance. This is another spot where it won't hurt to do too much.

Lowering the engine down at an angle, Cardines of Honda Fiend Racing gives it one final look to make sure the new engine isn't going to hit anything on its way down.

into the rear mount. Then attach the top transmission mount. Once all three mounts are attached, the rear engine bracket can be bolted to the engine from underneath. Then tighten up the left and right mounts.

Once installed properly, most B-series engines will fit under the hood with ample clearance on all sides of the engine and transmission. Certain LS/VTEC combinations and taller-deck B18C engines will necessitate minor amounts of trimming to the underside of the hood to avoid rubbing. You can do this using a small grinder to cut small portions of the webbing out from the underside of the hood. If the vehicle isn't too low to begin with, another solution is to space down the engine with a few hardened washers on the left and right engine-mount studs. Both are acceptable solutions for these all-too-common hood-clearance issues.

Axle and Suspension How-To

Depending on which model Civic is receiving the engine swap, putting the underside of the vehicle back together can be either easy, or a little bit tricky. For all models excluding the HF and STD, the factory-issued suspension components can be reinstalled without a hitch. For the HF and STD models, their smaller, non-Integra-like brake hubs will

These '88–'91 Civic EX knuckles, hubs, rotors, and calipers were installed for extra braking. This is another bolt-on procedure for the Civic.

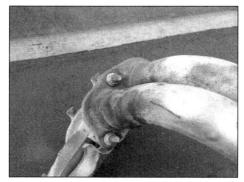

When selecting a B-series exhaust manifold, remember that almost all are compatible with any B engine. Be careful not to mix and match top halves and down pipes, as the later models are not compatible with the earlier. Try to find a complete unit.

need to be swapped out in order to accommodate the Integra axles. Although custom axles can be made for these hubs, they're ultimately too small in diameter and will break on a regular basis. Any of the other '88–'91 Civic front suspension knuckles and hubs will work perfectly. However, it's recommended to use the front knuckles and hubs found on the four-door EX model, as these house the largest disc brakes available for this chassis. If this knuckle and hub combo is chosen, it's imperative that the EX calipers, rotors, and brake lines be installed as well. These will all aid in better braking.

The axle installation will be similar for all Civic models at this point. Providing you're using the proper intermediate shaft from the '90–'93 Integra, and not the JDM piece, '90–'93 Integra axles can be easily installed. Although the JDM intermediate shaft is shorter, it may be used providing that custom-length axles are used. Before installing the axles, the inner dust ring must be pried off the inside of the hubs in order for the wheels to spin freely. Make sure to perform this task on both sides of the car; otherwise, you'll be taking things apart later.

A second driver side axle solution involves using the '94–'01 Integra intermediate shaft and axle. This works perfectly as long as the inside axle seal on the transmission is swapped out for the '94–'01 Integra seal. Failure to do so will result in transmission-fluid leakage. Although other axle and intermediate-shaft combinations are available, these have proven to be the easiest and most reliable. Some other axle combinations require the customizing of several axles just to produce one final unit. The nice thing about going the aforementioned route is the simplicity when it comes time to change a broken unit.

Shifters, Cables, and Such

In order to finish the underside of the car, the shift-linkage rods are the next. HCP Engineering, Place Racing, and Hasport Performance all manufacture pre-fit shift mechanisms for this transplant. This is by far the easiest route to take. Although it is possible to use a linkage set found on any of the

This custom-fabricated shift linkage from HCP Engineering will provide you with the simplest install. Notice the Integra-style ends in the lower portion of the photo.

'90–'01 Integras and to cut, modify, and weld it to fit, making it fit properly is a long, two-person process. When using the aftermarket linkage, simply install your old shifter and shift knob on the new rods, and you're ready to go.

After installing the shift mechanism, the original Civic clutch cable will need to be reattached to the new gearbox. A throttle cable from any '90–'93 Integra will complete the pedal assembly, and it installs in much the same way as the original Civic cable. Instead of using the Integra cable, HCP Engineering and Place Racing make throttle cable brackets that work in conjunction with the original Civic cable.

Fuel-Injection Guidelines

Moving on to the fluid systems, the fuel injection feed line is the next item to hook back up. Thanks to the similarities between all of the B-series fuel rails, in most cases, the original Civic fuel feed line will fit just is if it were meant to be there. If the Civic of choice does not use the large-damper pulsation fuel cap nut, then the swapping of fuel injection feed lines will be in order.

Civics that use the small fuel cap nut will need to have their fuel injection feed lines replaced with a unit from any CRX Si. Once installed, they'll all hook up to the B-series rail in the same manner. Moving on to the return side, you'll find that due to the larger fuel-pressure regulator outlet on certain B-series engines, the Civic's smaller fuel-injection line may need to be modified to fit. If needed, a hose reducer may be inserted into the original line, which will allow it to step up to the larger size regulator. Be sure to use the proper fuel-injection hose and clamps to finish things off properly.

Cooling Concerns

Regardless of which model Civic is being used, the original radiator will fit right back into place with the cooling fan. On all B-series engines, the Civic's heater hose may be reattached to the new cylinder head's water inlet pipe without any cutting.

Depending on which engine is being installed, the upper radiator hose selection will vary. Non-VTEC B-series engines can use any Integra RS, LS, or GS upper hose, while the late-model GSR engines must use one from a '94–'01 Integra GSR. All B16As, B17A1s, and Type R engines can use an upper radiator hose from a '97–'01 Integra Type R or any Del Sol with the DOHC VTEC.

Lower radiator hoses from any vehicle originally equipped with a B-series engine will fit. However, the late-model GSR hose is the longest. The radiator necks are smaller in diameter than the Integra hose, so be sure to use adjustable-tension hose clamps. Most aftermarket aluminum radiators provide additional cooling and have larger necks that will solve this problem. Another option is to install a Del Sol DOHC VTEC twin-core radiator with the larger inlets and outlets. Although it won't bolt up into place on this chassis, Place Racing makes a special lightweight crossmember that will make installation a cinch.

Air Conditioning Guidelines

A/C components are just as easy to retain here as with other Civic to B-series swaps. With the addition of a special A/C bracket from any of the motor-mount manufacturers mentioned earlier, you could bolt up the original Civic compressor to the B-series block with minimal effort.

It's important to note that there are two different A/C compressors in circu-

This specially engineered air-conditioning bracket from HCP Engineering allows you to reuse the Civic air compressor and lines. Two different versions are available, depending on which compressor you have.

You might have to remove the cap nut if it isn't compatible with your particular fuel injection feed line.

HCP Engineering engine mount kits leave you plenty of room for the OEM radiator and an aftermarket header.

If you're using the HCP Engineering bracket when reinstalling you're A/C, keep the original pulley and bolt because they will be reused.

lation for the '88–'91 chassis, both with entirely different bolt patterns. These are referred to as either Matsushita or Sanden units. It's imperative to identify which one you have before ordering the necessary bracket.

Since all of the original lines can be reused, you can actually leave the compressor in the car, tucked out of the way during the entire swap process. Going this route will eliminate the need to have the air conditioning system refilled once the car is up and running.

Maintaining Power Steering

Like the A/C, retaining power steering won't be too difficult, providing you have the right parts to begin with. With only a handful of fourth-generation Civics being equipped with power steering, it's not the most common thing to hook up on this particular swap. Since the CRX and several Civic models weren't adorned with power steering, that leaves only the LX, DX and Wagon models to contend with. Due to the differences in the Civic-style power steering pump and

the B-series brackets, a pump swap will be necessary. To make it work, you need the pump that corresponds with the engine.

Since both the upper and lower power-steering brackets are usually left on the donor engine at the junkyard, this will be a straight bolt-on procedure. In order to hook up the power-steering feed line originating from the steering rack, the fitting on the end of the Civic line must be cut off and replaced. Depending on which new pump you use, you must fit the corresponding power-steering-line

Air Conditioning Issues

With so many different air conditioning compressors and compressor brackets, it can sometimes be just downright confusing to choose the correct parts. Honda not only manufactures different compressors for each of its different vehicles, but also for each major body style and engine change. With all of these variables, we're talking about more than a dozen different compressors and just as many different brackets. Those numbers are enough to make life very confusing, to say the least.

Most of the big-name aftermarket engine-mount companies now manufacture their own A/C brackets designed specifically for different engine swaps. If you're looking to add or retain the air in your '94–'01 Integra or '92–'00 Civic with an H-series transplant, then companies such as Place Racing, Hasport Performance, and HCP Engineering have got you covered with their patented brackets.

Another popular engine swap in which the A/C is often retained is the '88–'91 Civic B-series transplant. Once again, the aftermarket has come to the

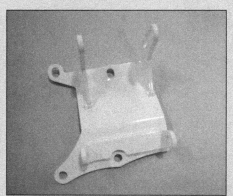

This bracket is designed for the use of an '86–'89 Accord compressor mated to a B series engine.

rescue. In this case, all of the three major engine-mount companies manufacture brackets that get the job done equally well. Just make sure that you use the belt suggested by the bracket manufacturer. This is important since the stock belt may be the incorrect length. It's also important to note that on many vehicles, Honda produces more than one type of A/C compressor. Take note of which one your car has before ordering your bracket, or it may not bolt up. Usually the compressor manufacturer's name is located on top of the unit in the form of a sticker or an engraving.

With most of the other swaps that you'll read about in this book, you'll be able to hook up the A/C with all genuine Honda parts. For example, any B-series swap into the '92–'00 Civics can simply use the bracket found on any DOHC VTEC Del Sol.

This bracket is special in that it allows you to mate the Civic A/C compressor with any Integra-style engine block. Keep in mind that on these two chassis, the A/C bracket also serves as the front left motor-mount bracket. Interestingly enough, the DOHC VTEC Del Sol was the only car Honda offered with a B-series engine and the Civic A/C compressor. Not even the similar '99–'00 Civic Si has this bracket.

Most of the other swaps in this book that allow room for A/C will use parts from your donor engine. Civics with ZC or D-series swaps, Accords and Preludes with H-series swaps, and Integras with B-series swaps will all generally require little effort to find the proper bracket and compressor combination. In all cases, check first to see if the original bracket from your old engine bolts to the new engine. If not, then confirm whether the donor engine's bracket bolts to the original A/C compressor. In most cases, one of these two scenarios will prove to be a winner. If not, then you'll need to look to the aftermarket brackets just mentioned, or perhaps fabricate one yourself.

and have it brazed or TIG welded to the end of the Civic line that has been chopped off. Once completed and leak free, the power steering will function properly. Although a custom line can be made, it is usually a more-expensive route to take.

Post-Swap Recommendations

After installing one of the more-powerful B-series engines, upgrading of the braking system is a very good idea. Following brake upgrades from the EX mentioned in the introduction of the '88–'91 Civic section, you'll surely have plenty of stopping power for any situation that you may encounter. Apart from braking, the extra 100 pounds from the B-series engine won't pose any major suspension problems. In fact, several road-race teams have chosen the B-series engine despite its additional weight. With that in mind, you can expect that the extra weight won't throw off the suspension too much for your street car. If you're still concerned about issues related to understeer, Progress Suspension's stiffer rear anti-sway bar and specially designed lowering springs might be in order.

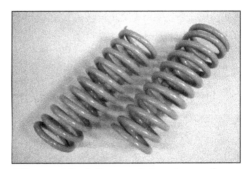

When you install a much heavier engine, you might want to consider these specially designed engine-swap springs from Progress suspension. They lower the vehicle and counteract understeer, which make them a must-have.

A Final Word for the B Series

As far as reliability issues are concerned, other than the all-too-noticeable vibration from some urethane engine mounts, the ED/EE to B-series transplant will easily provide you with years

The tightest fit you'll find is putting the B series GSR into the '88–'91 Civic chassis; it leaves little room left over. Nevertheless, the GSR is also among the most powerful B engines to choose from.

of reliable enjoyment. Since the Honda DOHC VTEC engines are some of the most well-built and reliable engines in the world, it would be fair to assume that this entire swap could be considered one of the most drivable and reliable swaps out there. Even at twice the cost in comparison to the ZC or the SOHC VTEC engine swaps, for those interested in speed, the benefits of the B series will most likely outweigh the money spent.

H-Series Engine Swap

More than likely, you aren't going to find too many '88–'91 Civics with an H-series engine packed under the hood. The reason for this is simply the level of difficulty and unavailability of parts for this particular transplant. Fortunately, for those of you interested in such a swap, manufacturers such as Place Racing have contributed a substantial amount of research and the proper components to make it work. Although you may only see cars such as these on the racetrack, with the help of the aftermarket, more will be spotted on the streets soon enough. This is one of the most difficult and time consuming of transplants in this book and isn't suggested for the novice or the first-time engine swapper. An experienced mechanic with all of the proper tools and parts could complete this swap in approximately two days. For the amateur, it could take significantly longer, if it even got finished at all.

This H22A Prelude VTEC engine is just waiting to be installed. It is much easier to work on a chassis this small without the exhaust manifold in place.

H-Series Engines and Transmissions

Now that you've read that cautionary warning, let's assume you're up to the task. Although this is one of the largest displacement four-cylinder engines that Honda has to offer, you might be surprised at just how many different Prelude and Accord engines you can stuff into your fourth-generation Civic chassis. You will find that just about any H- or F-series engine will fit. Although the list can be quite lengthy, several H-series engines exist around the world. They can be found not only in the United States, but also in Japan and various portions of Europe. Among the most popular H-series candidates for the ED/EE-chassis Civic are the USDM and JDM H22A engines. With 200 horsepower from the JDM version,

Although the F-series SOHC Accord engines are physically the same as the H series as far as mounting points are concerned, installing one of these in a Civic would be rather pointless.

these engines make a very attractive addition to the '88–'91 Civics.

Many other H22As can be found in the likes of the EDM Accord Type R, the USDM Prelude SH, and the JDM Prelude Spec S, just to name a few. If a non-VTEC H series is what you're after, look no further than the 160-horsepower '92–'96 USDM Prelude Si. Popular F-series engines that may fit into this chassis include the powerful but non-VTEC F20A from the '90–'93 JDM Accord 2.0 Si and any other F-series engine, SOHC or DOHC. Not that you'd want to swap in a SOHC Accord engine, though.

Transmissions of choice for these swaps include any '92–'96 Prelude units: USDM, JDM, or EDM. Late-model '90–'97 hydraulic Accord transmissions will also bolt up to the engine but differ slightly in their mounting point on top of the case for the right-side engine mount. It is possible to modify these cases to fit with the proper drill and tap set, but this usually isn't necessary since the gear ratios on the H-series transmissions are far superior to begin with. The '97 and newer ATTS (active torque transfer system) transmissions should be avoided because the required computer is difficult enough to wire up on a later-model car, not to mention an older, OBD 0 Civic. Regardless of which engine or transmission combination you decide to go with, with the exception of the high-performance Spec S and Type R engines, all of these engines and transmissions may be found in abundance at most wrecking yards for prices surprisingly lower than most B-series engines.

ECUs and Wiring For the H Series

In most cases, your best bet for an ECU when doing a DOHC VTEC swap is the P13. These can be found in several JDM Preludes with the H22A engines, as well as the '93–'95 USDM Prelude Si VTEC. OBD II ECUs should be avoided for H swaps into the '88–'91 chassis unless you're using the Accord Type R engine and ECU. Even in the case of the Type R, both the P13 and the P72 GSR computers are known to work extremely well. Non-VTEC H-series swaps will all require the P14 ECU found in the

'92–'95 USDM Prelude Si. In all cases, if an OBD II Prelude unit is to be used, avoid all USDM computers. Anti-theft systems render these newer ECUs worthless once removed from their vehicle of origin, not to mention the USDM emissions issues. OBD II JDM ECUs are all acceptable if OBD II is required.

Since all '88–'91 Civics were manufactured before the introduction of OBD, and all H-series engines are either OBD I or OBD II, there is going to be some electrical work involved. With an OBD computer, an underdash adapter will need to be retrofitted into the vehicle harness to accommodate the newer ECU. Since the plugs differ on the two versions of OBD ECUs, it's important to identify the proper connectors before soldering them onto the vehicle wiring harness. Once that's completed, the original Civic engine harness must be retrofitted to accommodate the donor engine into the car's engine bay.

On models only equipped with two fuel injectors, you'll first need to add two additional injectors into the engine harness. Another option is to start fresh

On the dual-injected Civics, two additional injectors must be added inside the engine wiring harness of the vehicle receiving the swap. The new engine will in no doubt be equipped with four injectors.

with a multi-point-injection harness from the EX, Si, or HF models. Either way, Prelude-style injector plugs must also be used to replace the Civic plugs on the engine harness. When using Prelude-style injectors, a resistor box must be used. These can be found on many Preludes, as well as multi-point-injected Civics and Integras.

Further wiring modifications include hooking up the exhaust gas recirculation (EGR) valve and the intake air bypass sensor. These modifications are manda-

This EGR valve is strictly an emissions-related device and must be wired up to avoid a check-engine light. Unlike other check-engine lights, when it's flashing this code, it won't affect performance.

tory if you use the P13 or P14 Prelude ECUs. Those using P72 GSR or P30 Del Sol ECUs can skip these two steps. For the Civic harness to fit properly on the new engine, certain wires will need to be cut and extended. All harnesses will require the Prelude's alternator wiring, including the plastic cover that sits on top of the engine. Grafting this into place will provide you with the necessary connections to hook up the harness to the Prelude alternator. The oil pressure switch, reverse light connector, distributor plugs, oxygen sensor, and fan switch connections must all also be swapped out for plugs found on the donor engine's harness.

Some Prelude engines use an external ignition coil setup that must be

Grafting the OEM Prelude alternator wiring onto the Civic harness will provide you with the most stock-looking appearance. However, that may be the only thing under the hood that looks stock.

addressed upon modifying the wiring harness. If you have one of these distributors, there are two choices for you. You may retain the external coil by installing either the Prelude coil or an aftermarket coil on the frame along. Don't forget the plug wire to go with the aftermarket coil. The second solution involves converting the distributor to use an internal coil. Either solution is satisfactory.

To finish up the standard wiring, a knock sensor found on all Prelude engines must also be accounted for when modifying the engine harness. Once that's done, VTEC can be wired up if needed. All VTEC engines will require the addition of wires for the VTEC solenoid, and in most cases, the VTEC pressure switch will need to be wired into place as well. Certain JDM H22A engines aren't equipped with a VTEC pressure switch, and so it won't be required. This is only true if the JDM ECU is also used. If an older ECU is being used, the computer will throw a malfunction code for that missing sensor, thus disabling VTEC. If an older computer is desired, the VTEC solenoid assembly must be swapped out for an older unit with provisions for the pressure switch.

Installing the H Series

The number one problem with the '88–'91 Civic chassis and the H series motors is space; or more appropriately, the lack thereof. It's tough enough trying to shoehorn a B-series engine into place in an EF Civic, let alone a larger and heavier H22A. Apart from this major obstacle, you're not going to run into any problems that you wouldn't normally run into when swapping the H- or F-series engines into any other type of Civic. Thanks to the folks at Place Racing, a complete mount kit is now available that will enable you to place any of the H- or F-series engines into any '88–'91 Civic. Unlike any of the other mount kits on the market, welding is required, but then again, what else would you expect when trying to stuff a motor this size into a car this small?

Place Racing's engine mount kit consists of three special brackets designed to be welded onto the frame.

With the engine removed and the engine bay free of clutter, you need to remove both the left and right side pocket mounts from the frame rails.

Place Racing provides a special drill bit for drilling out the spot welds on the frame. This results in a clean, undamaged surface for the new mount. Before welding on the new pocket mounts, remove paint and rust from the frame in order to ensure a proper weld bead. Next, the pocket mounts may be welded to the frame using the supplied templates from Place Racing. You can either MIG or TIG weld them, depending on the desired final appearance. Once welded securely into position and cooled down, apply a couple of coats of automotive-grade spray paint onto the mounts so they won't rust in the future.

In addition to the two left and right frame-rail brackets, the rear mount must

This pocket mount in the middle right of the photo must be drilled out and removed before installing the specialized bracket from Place Racing.

Remove the factory spot welds in a clean and quick manner using this special drill bit from Place Racing.

also be taken into consideration before installing the new engine. A special bracket from Place Racing allows you to attach the factory Prelude rear mount.

Once the original Civic rear rubber motor mount is removed, the paint surrounding this area will need to be sanded away. Using the supplied template, weld the triangular-shaped bracket into position on top of the rear crossmember. Once the weld has been painted, a '92–'96 Prelude rear engine mount (from a car with a manual transmission) can be bolted into place. You can use a driver-side engine mount from any '92–'96 Prelude, but the top transmission mount must from a manual transmission model.

With all of the mounts unattached minus the rear mount, the engine should now be lowered into position at a downward angle. Before the engine is pushed back all the way toward the firewall, the

Many folks choose to manufacture their own custom mounts in lieu of the Place Racing kit. Either way, welding is going to be required.

This custom transmission mount is a combination of a Prelude bracket and a specially fabricated steel mount. Together, they hold the Prelude transmission into place.

rear engine bracket should be slid into place. Do not bolt it down just yet. Position both the left and right mounts into their respective pockets and install the necessary hardware. Next, remove the engine hoist and fasten the rear bracket. Then, tighten down all three mounts to the proper torque specifications provided by Place Racing.

Although it's an extremely tight fit, thanks to the engineers at Place Racing, you can miraculously fit a DOHC Prelude engine into this bay without hitting the hood or touching the pavement below. In order to ensure that hood damage doesn't occur, it is recommended to trim off the underlying hood skin in the areas near the valve cover. This skeleton of metal reinforces the hood and adds to its rigidity, so you won't want to remove the entire skeleton. Cutting out a small portion in order to allow extra clearance won't pose any problems to the hood itself.

Alternator clearance is at a minimum and must be accounted for once you install the engine. It's necessary to cut the headlight housing slightly so it won't hit.

Final clearance issues will need to be addressed concerning the left headlight. The housing for the driver-side headlight must be trimmed in order to accommodate the H-series alternator, which sits in front of the engine block. Wait until the engine is installed to perform the trimming so that the proper measurements can be made.

Axle and Suspension Protocol

Once the engine is tightened into place, the underside of the vehicle should be finished off. HF or STD vehicles need a bit of work on the suspension. Due to

the smaller diameter of the wheel hub housings, the larger-diameter axles won't be able to slide into place. Although custom axles may be made using the smaller HF and STD shafts, these will break in no time. Swapping the knuckles, wheel hubs, brake lines, calipers, and rotors of the '88–'91 EX model Civic will be your best bet. Although other Civic knuckles will transfer over and work, the EX has the largest brakes of the bunch, and they'll prove necessary with the bigger H engine.

Before tightening up the new suspension, the proper axles should be installed into the transmission and wheel hubs first. Axles for this swap can be ordered through swap companies including Place Racing. If you're inclined to take the measurements yourself, then any competent driveshaft rebuilding shop should be able to prepare a set for you in no time at all. When having custom axles built, any of the Prelude or Accord intermediate shafts may be used. Just be sure to specify to your prospective axle supplier or builder which one you intend to use. Also, be sure to remove the dust rings on the center of the wheel hubs to ensure that the wheels will spin freely when you're done.

To finish things up underneath, a custom crossmember from Place Racing will need to be installed to allow frontal clearance for the larger engine. Once installed, the new crossmember has provisions for the '92–'00 Civic radiator, which is significantly smaller and will free up lots of space.

Beware here, as Place Racing makes two different front crossmembers for

Both of these crossmembers are designed for the '88–'91 Civic from Place Racing. The unit on the right is much more stout and is designed specially for swaps with an H-series engine.

the EF chassis. One is designed for the H-series swap, while the other is designed for the original engine, the ZC, or the B series. The lower radius rods custom-made by Place Racing connect to the front crossmember. These must be replaced, as the original driver-side radius rod will interfere with the H-series crankshaft pulley. Once again, the use of the proper crossmember is mandatory, as these special radius rods won't work on the wrong unit.

You need to install the Place Racing radius rods in conjunction with their crossmember. The OEM crossmember and rods will hit the crankshaft pulley.

Prelude Shifting Made Easy

Moving on to cables and such, the shifter mechanism will be next. Since all newer Preludes and Accords use shifter cables in lieu of shifter rods, the Civic shifter rod assembly can be tossed. A cable assembly from any '92–'01 Prelude or any '90–'97 Accord will suffice, although you have to use the complete setup from either the Prelude or the Accord.

When using the special Place Racing shifter mount box, the Prelude or Accord cables can be removed from their original housing and attached to the new box. The unit must be bolted to the underside of the chassis, allowing the cables to be routed completely underneath the vehicle. For those folks who choose not to use this handy mounting box, the original Prelude or Accord shifter plate may be bolted to the inside of the vehicle from up top. In either case, before bolting the plate or the box into place, the material under-

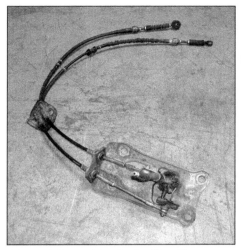

Many shifter cables look alike. These were purchased as a unit from a local wrecking yard. They came from a '92–'96 Prelude with the manual transmission.

This cable-to-hydraulic actuator kit is a must-have for all H-series swaps into the '88–'91 Civic chassis. Notice how easily it bolts into place, allowing you to reuse the OEM clutch cable.

neath where the shifter will sit must be cut out with a reciprocating saw. Cut just enough to allow the shifter to move in all directions, while still leaving enough material to bolt the plate to.

When using the factory plate, take four 8-mm bolts with nylock nuts and attach the Prelude or Accord assembly to the chassis. When going this route, the cable will run inside the vehicle and exit through a 2-inch hole that you'll need to drill out 18 inches in front of the shift lever. At this point, whether you use the Place Racing kit or not, the cables will be routed to the transmission the same way. Under the chassis, the cables should be directed over the rear engine mount toward the passenger side. Once near the gearbox, they can be attached using the proper washers and cotter pins. When installed properly, the shifter should freely engage all gears. If it doesn't, double-check the hole to verify that the shifter isn't hitting the frame.

Adapting To Hydro

Since the Prelude and Accord transmissions are only available with the hydraulic clutch operating system, the use of a special adapter bracket from Place Racing is mandatory. Although you may adapt the vehicle to use a hydraulic transmission in other ways, this would require another pedal assem-

bly, a hydraulic master cylinder, a clutch fluid reservoir, and some hydraulic lines.

Using this timesaving piece from Place Racing will enable you to reuse the clutch cable from the Civic by adapting it into a linear movement to activate the clutch-release arm. This invention cuts down hours of labor and saves you from acquiring an expensive hydraulic clutch system.

Once bolted to the transmission case, the cable can be installed, thus completing the install. The original throttle cable may be reused providing that a Place Racing adapter bracket is installed on the manifold. The adapter is necessary because the original Civic throttle cable cannot be reattached to the Prelude bracket.

Fuel Injection How-To

The fuel system should be addressed next. On the easy side of the fuel system, the return line will slide onto the Prelude regulator provided you use a longer piece of fuel line. The feed side won't be so straightforward. First off, the fuel inlet is on the opposite side of the engine bay with certain Prelude engines. A rail from the '92–'96 USDM H engines will provide a quick solution. Regarding the fuel injection feed line: the Civic is available with two different ones, depending on the model. Si models that use the large damper pulsation nut

will require their fuel injection feed lines to be swapped out for the versions that use the smaller retaining nut. These can be found on most other Civics of this era. When finished, always ensure that the proper aluminum seals have been replaced and that there are no fuel leaks before driving the vehicle.

Retaining the Amenities

From the fuel system to the cooling system, once again you'll be using a hodgepodge of Civic and Prelude components. Use a radiator from either the '94–'97 Del Sol DOHC VTEC or the '99–'00 Civic Si. Since the Place Racing crossmember doesn't have any provisions for the wider EF-style radiator on the new crossmember, the newer and smaller version must be used. The radiators found in the two vehicles above are recommended because they're twin-core pieces. They not only have superior cooling capabilities, they also have the larger inlets and outlets, much like you'll find on the Integra or the Prelude. Even better, you might want to consider an aftermarket unit from the likes of Koyo or Fluidyne.

Whichever unit you choose, a thinner radiator fan is always necessary for this swap. The fan must be mounted to the front side of the radiator, just behind the bumper opening. This will keep it well out of the way of the slave cylinder on the front of the transmission. Be sure you purchase a fan that is able to push

A smaller fan such as this will be mandatory because of clearance issues. You'll find that you can mount these on either side of the radiator since they can either pull or push.

and pull depending on how it is mounted. Be sure to wire it up so that it spins in the proper direction.

As far as cooling hoses are concerned, upper and lower radiator hoses from any '92–'96 Prelude will suffice. If the heater is to be retained, the hose exiting the rear main water pipe on the Prelude engine must be attached to the water valve on the vehicle's firewall. A section of heater hose with adjustable hose clamps will be necessary for this to work. The opposite hose exiting the other side of the water valve will slip onto the water fitting on the side of the cylinder head inlet. Since the Civic hose is a bit too short, you'll want to purchase one from any '92–'96 Prelude for a perfect fit.

With the radiator of choice in place, the hoses hooked up and the fan fastened on, it is clear that there is no room for any type of air conditioning or power steering system to be installed under this hood. Keep this in mind before doing any of these transplants: everything is a tight fit.

The H Isn't For Everyone

Apart from the difficulties associated with the installation process, things aren't going to get any better when dealing with cost issues. This point can be illustrated by looking at the initial costs of an H- or F-series engine swap. The difference in cost between swapping these engines into the ED/EE chassis and swapping them into a later-model Civic is quite consider-

able. Many aftermarket components are needed to make this engine function properly in a vehicle as old as this generation of Civics, which no doubt means more cash.

A successful swap is measured by its reliability and drivability. When dealing with the '88–'91 chassis and the H- or F-series engines, it's sometimes rather difficult to produce a finished product able to meet both of these criteria. For example, as with most H-series swaps into Civics and Integras, the axles are always going to be an issue. Whether it's torn constant velocity (CV) joints or a snapped axle shaft, axle gremlins on some vehicles seem to never go away. Other adverse effects of this swap include an increased amount of understeer due to the additional weight added to the front of the vehicle. This can be overlooked, however, if you don't plan on doing any autocrossing or road course racing.

The bottom line is that these particular transplants are best suited for those who put a greater importance on straight-line performance than on reliability or drivability. With major engine vibrations, excessive noise inside the vehicle, inferior braking, and an excess of torque that the front tires can't hold, it's easy to conclude that everyday drivability is rather unrealistic. With that said, it never seems to stop those diehard enthusiasts willing to make sacrifices for the sake of speed. So, you might be asking yourself, with not-so-perfect drivability characteristics, why would anyone even want to do one of these swaps after reading this?

Well, it usually reverts to the old power versus weight ratio. The EF chassis is among the lightest in Honda's stable, and adding an H- or F-series engine will produce an extremely quick car. That alone is good enough of a reason to do a swap for most of the customers I've dealt with in the past. In the end, a 1,800-pound chassis with anything over 200 horsepower is a sure-fire recipe for a rocket.

Coil-overs from the folks at Progress Suspension will allow you to fine-tune the front and the rear of the vehicle to suit your needs. This is helpful when you install the much heavier Prelude engine.

This swap is best suited for a drag car. You'll be hard pressed to get one of these to pass a smog check legally. The guys at Honda Fiend use this CRX strictly for racing.

1990 TO 1993 INTEGRA

The second-generation DA/DB Acura Integra received its fair share of aftermarket support right from the get-go, as it was the first Honda or Acura vehicle factory equipped with a 1.8-liter engine. Developed for the United States in 1990, the DA/DB chassis Integra remained unaltered in engine or body style until its replacement by the third-generation Integras in 1994. For four years this Integra battled it out at the dealer lots with the likes of the Nissan 240SX and Mitsubishi Eclipse. With only 140 horsepower available from the most powerful B18A1s, the turbocharged Mitsubishi Eclipse and rear-wheel-drive Nissan 240SX made for some stiff competition.

Unlike its rivals, but like most Hondas, the Integra's sales remained steady. With many of these Integras up for sale in the used car market, they make for excellent engine swap candidates for those wishing to step up to the Acura name without the high cost of the newer Integra or RSX platforms.

Integra Offerings

The '90–'93 Integras only have a few trim levels. Integra's levels of luxury begin with the RS, followed by the LS, and end with the top-of-the line GS. In '92, Acura introduced the GSR model, which was equipped with a more powerful engine and added amenities including a leather interior. The Integras were not the lightest vehicles of their time, and the Integra GSR weighed in at over 2,650 pounds. The GSR engine made up for this significant weight gain over previous Integras, and was the most powerful

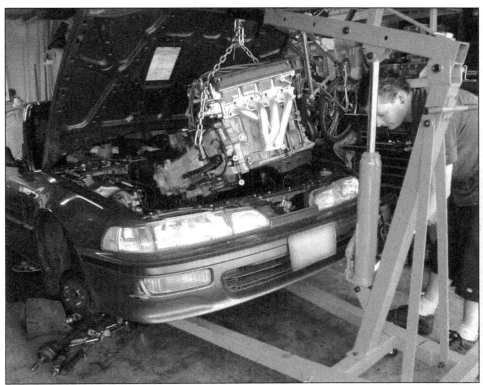

B-series swaps into the second-generation Integras are among the easiest. Robert Young of Holeshot Racing lowers this B16A in after just a couple hours of work.

This base-model Integra features the 140-horsepower B18A engine. Although not the most powerful, it is at least a DOHC and worthy of upgrades.

Vehicle ID Numbers and Engine Swaps

The vehicle identification number (VIN) is a unique compilation of numbers and letters assigned to every vehicle by the manufacturer. Since its implementation, the 17-digit VIN has served its purpose mainly for vehicle registration and identification. When interpreted properly, each character of the VIN will identify specific information regarding the vehicle. Among its many functions, the VIN is interpreted during emissions testing and compared against the engine underneath the hood. In some states, following an engine transplant, the proper paperwork must be filed with the state smog referee to avoid future emissions testing problems. The VIN may be located on the firewall, the driver-side corner of the dash, or driver-side doorjamb. The simplified chart on the right shows what each character of the vehicle identification number represents.

Sample VIN: 2HGEH234XNH537228

Digit	Meaning
1	nation of origin
2	manufacturer
3	vehicle type
4, 5	body type ('83–'86 years)
4–6	chassis and engine ('87–'98 years)
6	transmission ('83–'86 years)
7	body type ('83–'86 years)
7	body type and transmission ('87–'98 years)
8	trim level
9	check digit
10	model year
11	assembly plant
12–17	production number

This is the easiest place to view the VIN.

Honda four-cylinder to make its way into the United States so far.

Civic engine swaps in the early '90s were few and far between, so the Integra made the most sense to those few Hondaphiles interested in going fast. It was clear to folks by now that the B-series engines were necessary for big horsepower. This was not only recognized by the public, but by the aftermarket as well.

Except for the Integra's standard B-series engine, all models were equipped with double wishbone suspension and four-wheel disc brakes. The GSR braking system is a bit stouter in keeping with its high-performance theme. Second-generation Integra owners who wish to upgrade their braking systems need not install much more than a set of slotted brake rotors and high-performance pads. Suspension components from the likes of Tokico and Progress Suspension are available to fit all types of racing use.

If you pop the hood on any model except the GSR, you'll find an engine stamped B18A1. With only 130 to 140 horsepower (depending on the year), these Integras were not considered very fast, but with the right upgrades they sure could be. Those willing to step up and spend the extra cash on the GSR received the B17A1. This GSR engine can best be described as a de-stroked B18A1 with a B16A VTEC cylinder head, and produced 160 horsepower in its stock form. With this type of power, an engine swap on the GSR would be a foolish procedure to say the least. Although all the B18A1 and B17A1 engines receive plenty of aftermarket support, the B18A1 is lacking VTEC. Integras equipped with the B18A1 make the most sensible of candidates for an engine transplant.

Early OBD Issues

Between the different Integra models, two different OBD scenarios must first be discussed. Since '90–'91 DA/DB Integras production are OBD 0, it would make perfect sense to install an OBD 0 engine for the sake of simplicity. OBD 0 VTEC B-series engines feature computers that will plug in to the OBD 0 Integra's vehicle harness, without any modifications. Although OBD engines may be installed into these years of vehicles, if it isn't necessary, then it's recommended not to. Cases that warrant an OBD conversion include the late-model GSR, Type R, or H-series engine transplants. Since these engines are only being offered in OBD status, a vehicle harness conversion will be necessary to use these newer-style

Underneath the hood you'll notice that this early Integra features the OBD I electronics system. Since the Integras switched to OBD I in 1992, the other half of this body style are OBD 0.

computers. Since the OBD-style ECU plugs are different, they'll have to be retrofitted onto the Integra's vehicle harness so you can plug in the ECU.

The Integra computer is located under the carpeting on the passenger-side floorboard. The old ECU plugs can be cut off and thrown away. New OBD-style connectors must be acquired to match the ECU. Be sure to leave plenty of wire connected to the plugs, as you'll need it later when making the wiring connections. OBD I ECU users may find plugs off of any '92–'95 Honda or Acura vehicle. OBD II users will find theirs on the '96–'01 models. Due to differences in certain OBD II plug types, you may want to bring the new ECU along with you to make sure you get the proper plugs. This is important because of the differences between Civics and Integras in '96–'98 and '99–'01 respectively.

Once the proper plugs are gathered, the wires must be soldered onto the vehicle harness one by one. Using two separate service manuals (that of the Integra and that of the ECU), the proper connections can be made for the new ECU to be plugged in. Although OBD II ECUs can be wired into place, it's not recommended. If need be, the JDM units are your best option due to their lack of several emissions sensors found on the USDM versions. With any OBD conversion on the '90–'91 chassis, be sure to add the appropriate wires so that the malfunction indicator light will work. Since you cannot read engine trouble codes on the face of OBD computers, modifications must be made so they may be viewed on the gauge cluster.

Later OBD Issues

'92–'93 Integra owners will have a little different situation. These folks will want to avoid the OBD 0 engines and computers altogether. Since they're incompatible as far as ECU plugs are concerned, converting the vehicle harness to OBD 0 status would be considered a downgrade. With several OBD computers available that will work perfectly, an OBD I or OBD II engine should be your number one priority. If an OBD II engine and computer are selected, an adapter harness will need to

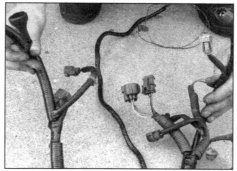

As with any swap, reusing the original wiring harness will always make the most sense. Almost all of the plugs and connectors are there, they just need to be rearranged and lengthened to fit.

be procured in much the same way as you would for the '90–'91 models. Again, avoid any of the USDM OBD II ECUs, as these will require major amounts of unnecessary labor.

Wiring Harness Rules

Regardless of which type of engine and computer combination is used, the original Integra engine harness should always be retained. You'll want to save the engine harness off of the donor engine for special plugs and connectors, but for the most part, it can be considered trash. On most swaps, the engine wiring harness off the DA/DB will need to be lengthened in various portions in order to accommodate various sensors on the new engine. Begin with the injector plugs to give you a solid starting point as to what plugs must be cut and lengthened. Remember to label the wires when cutting connectors and adding lengths of wire so that they don't get mixed up.

B-Series Engine Swap

Although the second generation of Acura Integras have already been equipped with a B-series engine from the factory, most Honda freaks will agree that there is always room for improvement. Every RS, LS, and GS Integra from '90 to '93 housed the B18A1, and the upgraded GSR model with the B17A1 was the only alternative. With well over a dozen variations of B-series engines available today, '90–'93 chassis Integra owners now have the

opportunity to put new life into their vehicle without a whole lot of difficulty. As you'd expect, most of these B-series engines will transplant into the '90–'93 Integra chassis with relative ease.

Thanks to the engineers at Honda, even after 13 years of production, the B-series engines are still much more alike than they are different. These similarities make this swap well suited for even the weekend mechanic. Most B-series swaps into second-generation Integra chassis could be completed in a matter of a short day by a competent engine-swap pro. The difficulty level is very low, assuming you have some mechanical knowledge and experience. Some B-series swaps that involve using the hydraulic-style transmission and advancing the OBD system will prove to be a bit more difficult. There'll be more on this later on.

B-Series Engines and Transmissions

Before choosing the proper engine and transmission, it's important to remember that Acura decided to upgrade the OBD 0 style electronics system to OBD I status for the '92 Integra. Although the body style underwent only very subtle changes, the entire electrical system underwent a complete redesign. Considering this, the '90–'91 models will require little to no wiring modifications if you stick with an OBD 0 swap, while the '92–'93 models will be easiest if an OBD I swap is used.

This late-model B16A engine has been mated to a cable-style transmission to simplify the installation process. Either the hydro or cable unit will both bolt up to the engine block in the same way.

With that being said, there are several B-series engines to choose that will add plenty of needed horsepower into this older Integra. Beginning with the B16A family, the '90–'93 JDM Integra RSi and XSi engines fit with ease, as do the '88–'91 JDM Civic and CRX SiR engines. Both of these B16A engines can be found easily and inexpensively in most junkyards. Remember on the Integra B16As to select an engine with the corresponding OBD system to avoid any unnecessary wiring modifications. You won't need to worry about this with the Civic B16As though, as they're mostly of OBD 0 status.

Finally, the B17A1 found in all '92–'93 USDM Integra GSR models, as mentioned earlier, may be swapped into the lower classes Integra chassis. With the B17A1 (like all '92–'93 Integra engines) being equipped only with the OBD I electrical system, installing it into a '90–'91 chassis using the corresponding P61 ECU will necessitate major wiring modifications. Of course, if a '92–'93 chassis is being used, then the engine will bolt right in with little wiring changes.

The same modifications can be performed on any late-model OBD I or OBD II B-series engine for that matter. With all of the higher-horsepower B-series engines produced in the mid- to late 1990s, sometimes OBD modifications are going to be unavoidable if you're using the '90–'91 chassis Integra. Without listing every late-model

B20B engines are popular alternatives for enthusiasts who seek the maximum in engine displacement. Later you can install a VTEC cylinder head and have the makings for quite a bit of power.

Cable-to-Hydro Conversions

When performing a late-model engine swap onto some of the older Honda chassis, sometimes the newer, low-mileage hydraulic transmissions can be more than tempting. What used to be a major ordeal involving a custom pedal assembly and hydraulic clutch system has now been simplified into one component. Cable-to-hydro conversions can be demystified with the use of a special adapter from the likes of Place Racing or Hasport Performance. By converting the cable movement into a linear motion that actuates the hydraulic clutch release arm, the changeover process has been reduced greatly. Along with the conversion bracket, which bolts directly onto the front of the transmission, a special mount from either of these companies will need to be installed in order to

Notice how the Place Racing conversion bracket bolts right to the front of the engine and reuses the original clutch cable.

bolt the transmission into place. As of now, these conversions are available for the B- and H-series engines.

B-series engine available, there are a few that often find their way under the hood of a second-generation Integra. These chosen few include the Integra B18 VTEC family, consisting of the GSR and Type R engines, as well as the non-VTEC 2.0-liter CRV engines.

Depending on the OBD system of the donor engine, wiring modifications will need to be performed on the vehicle harness in order to accommodate the newer-style computer. Standard additional wiring modifications will need to be addressed with the engine harness in order for all of the late-model electrical components to plug in properly. With the early style, OBD 0 PR3 and PW0 ECUs readily available from the B16As, converting the car to OBD I or OBD II status is rarely necessary, since these ECUs function quite well with the B18 VTEC also.

Since all of these '94 and newer B-series engines feature hydraulic-operated transmissions, the attachment of the cable-style gearbox from any of the B16A or B17A1 engines will be the easiest to swap. Not only are these cable-style transmissions much cheaper in price, they'll bolt right into place since they're from the same family of Integras.

If you're concerned about the price, you can even reuse your original B18A1 transmission if you don't mind the longer gears. If you're not concerned about price, then you might check out the JDM cable-style YS1 transmission equipped with an optional limited-slip differential. This is the crème de la crème of older-style Integra gearboxes, direct from Honda of Japan.

For those stubborn folks who must use the original late-model GSR or Type R parts, there are conversion kits avail-

Generally, limited-slip-differential (LSD) transmissions are labeled as such. If you want to know for sure, simply look inside the axle orifice on either side of the unit. On LSD-equipped transmissions, you can see straight through, while the standard units will have a shaft in the middle.

able from Place Racing and Hasport Performance that will allow the addition of these newer-style transmissions. Using the Integra's original clutch cable, these kits work in conjunction with a special actuator that releases the hydraulic clutch arm on the transmission. This is all done by means of a special bracket bolted to the front of the transmission case.

You'll need special engine mounts and brackets to use the later-model B-series engines. A special transmission mount will need to be used from Place Racing for the gearbox to be bolted up into place. With the late-model engines, be sure to remove all of the engine brackets off of the block, as none of these will be compatible with the older Integra's frame. If you use one of these later-style engines, simply swap the mounts over from the old B18A1 onto the donor engine.

This late-model B16A will fit perfectly reusing both the post mount and the engine mount of the B18A engine.

ECUs and Wiring for the B Series

As mentioned previously, there are a few different scenarios involving the electrical system, depending on which combination you go with. Speaking electrically, one of the easiest combinations is the '90-'91 chassis and any of the '88-'91 B16As. The JDM PR3 and PW0 ECUs will plug directly into the vehicle harness and allow you to simply add the VTEC wires. You'll need the VTEC pressure switch, the VTEC solenoid, and the knock sensor. Since the engine harness is already suited for a DOHC engine, you'll find that it will fit onto the B16As with minimal effort.

In addition, adding any of the '92-'95-style B-series engines to the '92-'93

Any OBD I ECU will plug directly into place when you're dealing with a '92-'93 version. Avoid using an OBD 0 computer, as that would be a bit of a downgrade.

chassis is an easy task. Although equipped with different mounts and transmissions, all of these engines have the same electrical configuration as the car. Any of the OBD I B-series ECUs will plug into the underdash harness of the second-generation Integra with ease. You'll also find that the original Integra's engine harness will readily plug into any of the aforementioned OBD I engines as well.

Depending on which engine and computer you use (and assuming you don't have a GSR model Integra), wiring modifications will include the additional wiring for the VTEC pressure switch, the VTEC solenoid, and the knock sensor for all VTEC engines. In addition, the B18C GSR engines will need to have wires added for the intake air bypass system and the evaporative purge control solenoid. Although the use of every OBD I VTEC ECU is very common on these swaps, it's important to note that if you wish to maintain the GSR's intake air bypass (IAB) system, then the P72 GSR computer must be used.

Lastly, for those of you who are simply looking to add the CRV engine without the addition of the VTEC cylinder head, the original B18A1 ECU may be retained. When using an OBD 0 ECU with an OBD engine, the electronics on the engine will need to be backdated in order to work. Simply swap the distributor, fuel injectors, and oxygen sensor for older B18A1 units.

A third and final wiring scenario involves the '90–'91 chassis and the '92–'95 engines. These swaps will require the modification of the wiring underneath

The wires for VTEC were just added. They exit the center of this harness and make their way upward. Running them as part of the original harness makes for a factory look.

the dash in order to plug in the OBD I ECU. The engine harness will also need to be modified, not only to add the VTEC components (if not using the CRV engine), but to use the new distributor, fuel injectors, and oxygen sensor as well. Although not terribly complex, this route can usually be avoided by using the PR3 or PW0 B16A ECUs in lieu of the late-model computers. This will eliminate the need to perform these additional wiring modifications to the wiring harness, yet still retain the VTEC system. When going this alternate route, the newer OBD distributor, fuel injectors, and oxygen sensor of the donor engine must be swapped out for OBD 0 components. Unless a Type R engine is being swapped into place, there really isn't a need to convert the car to OBD I or OBD II status, as the PR3 and PW0s are both excellent choices.

Installing the B Series

With semi-complex electrical issues behind you, you'll find that installing the engine will be a simple affair. I'm sure that you'd have figured that out already, seeing that this is an Integra chassis to begin with. Begin with the engine mounts. It doesn't matter which B-series engine you use, as all of the original manual-transmission Integra mounts will be reused.

This is, of course, if a cable-style transmission will be retained. As mentioned earlier, the later-model engines will require the brackets to be swapped out on the engine block as well. With the left mount attached to the frame and the original engine's bracket bolted to the block, the engine should be lowered into

This upper transmission mount from the Integra can be reused on the new transmission. Check it out thoroughly before you install it, because they have a tendency to break.

the bay with the transmission side at a slight downward angle.

The left engine mount can be attached at this time while raising the transmission side of the drivetrain. The right and rear brackets should then be attached and bolted into place. With the three mounts in position, the front mount can be set into place and tightened down along with the others. Once in position, you'll find that all of the B-series engines sit under the hood just like the old B18A1. It shouldn't surprise you that there is still plenty of hood and ground clearance, since all B-series engines have the same dimensions.

This one's a perfect fit with loads of room to spare. Notice how the intake, battery, cooling fan, and radiator all fit perfectly back into place. What else would you expect from another B series?

Everything Underneath

In keeping with the theme of simplicity, the original axles, suspension, and shift linkage may all be bolted up into their original locations. All of these components will be reused no matter which B-series engine or B-series cable-style transmission you use. With the cable-style transmission, the original Integra clutch cable may be reused and reattached to the new gearbox in the same way it was removed.

The original clutch cable may also be reused, even when swapping transmissions for a JDM unit. Simply slide the cable back into its original bracket, which was transferred over to the new engine.

Depending on which engine you choose, the throttle cable should be attached at this time. Cables found on the '97–'01 Integra Type R will work for any of these swaps excluding the GSR B18C. This swap will require the '94–'01 Integra GSR throttle cable. Due to the upside-down nature of the GSR's intake manifold, this will be a mandatory purchase.

Fuel and Water Systems

Moving on to the fuel system, you'll find that things are pretty straightforward to say the least. With all of the '90 and newer Integras equipped with the same style of fuel-injection feed line, the banjo and retaining nut will be sure to attach to the new fuel rail with ease. Be sure to replace the two aluminum crush washers with new ones in order to avoid any fuel leaks. Finish things off by slipping the existing fuel return hose over the new fuel regulator with the original clamp. As far as the fuel system goes, that's it.

Moving on to other fluid systems, thanks to the compatibility of all of the B series, you can retain the radiator, cooling fan, and lower radiator hose. Your choice of upper radiator hoses will be determined by which engine you use. All of the B16As, the B17A1, and the

The radiator may be reused in all cases and is perfectly capable of handling any cooling issues that the most powerful B series might have.

Type R engines will use the same upper radiator hose found in the '97–'01 USDM Integra Type R. GSR and non-VTEC engines will want to use the hose found on any '94–'01 Integra.

A/C and Power Steering Instructions

From cooling systems to staying cool, we'll talk about air-conditioning systems next. For those who choose not to sweat it out in the hot months, the ability to keep the A/C with these swaps is almost a no-brainer. The original bracket off the B18A1 block can be reused on any of the B-series engines. Simply bolt the original Integra air compressor up to the bracket and you're finished. It's hoped that none of the A/C lines were disconnected during the swap process, since they really don't need to be removed.

Other amenities (for those who

Every B engine will provide a factory fit. This B20B engine appears to have been there all along.

aren't concerned with weight reduction) that you might want to keep include the power steering system. Standard on all Integras, you'll soon see it's not going to be too difficult to maintain this luxury. The lower power steering bracket located on the original B18A1 engine can be removed and simply bolted onto the new engine. The upper bracket must be obtained from the appropriate vehicle if you're using a VTEC engine. Depending on which engine is installed, the upper power steering bracket found on the '92-'93 USDM Integra GSR may be used for almost all of these VTEC transplants. For transplants using the B18C GSR cylinder head, the appropriate bracket from any '94–'97 Integra GSR can be used.

Matching the proper upper power steering bracket to your pump is mandatory. If you are looking for one in a junkyard, note the P72 stamped on the front of this one.

Be careful not to purchase the bracket from the '98-'01 Integra GSR, as it is manufactured to fit an altogether different pump. With the correct brackets, the power steering pump will reattach in the same way that it was removed, using the original belt as well.

Braking and Handling

Since all Integras are equipped with four-wheel disc brakes, you won't need any additional stopping power when upgrading to B engines. Since all B-series engines weigh in at approximately the same weight, you'll also find the handling characteristics of the vehicle will remain unchanged. This isn't to say that Honda engineers haven't left any room for improvement in the braking and handling departments. With many aftermarket components available to

address these situations on the second-generation Integra chassis, significant improvements can be made. Just understand the fact that swapping in a slightly more powerful B-series engine won't require improvement in these areas.

B-Series Conclusion

Although the DA/DB chassis Integra to B-series engine conversion certainly walks the fine line of actually being considered a true engine swap, that doesn't make it any less of a viable option for those Integra owners who hunger for VTEC power. The nice thing about these transplants is that with the proper components, these engines will definitely feel at home underneath the hood. Utilizing many of the original Integra parts allows these swaps to be reliable, low-maintenance conversions. By reusing the Integra's original components, the costs associated with the B-series transplant can be kept to a minimum, while still yielding the utmost in performance and reliability.

Since this is a USDM engine and this vehicle is already OBD I, smog issues will be taken care of once this vehicle is certified.

H-Series Engine Swap

Those in search of more displacement for their second-generation Integra need not look any further than the H series. With the bottom-of-the-line stock H22A producing horsepower equivalent to that of a heavily modified B18A1, the decision to swap shouldn't be too difficult. Up until recently, H-series engine swaps into this chassis had involved quite a bit of fabrication and welding. Thanks to the folks at Place Racing, the H-series transplant

has become a true bolt-in affair. Using all of the proper components, a typical timeframe on such a swap should be around two to three days. The H-series units are much larger and heavier, and a little bit more difficult to shoehorn into any engine bay other than what it was designated for originally. Although the wiring portion of the swap will prove to be quite extensive in certain cases, the majority of the transplant process is straightforward. Still, it is suggested for those with at least intermediate mechanical skills.

H-Series Engines and Transmissions

Since this particular Integra chassis falls into two different OBD categories, vehicle selection, as opposed to donor-engine selection, is more important. All the H-series engines are OBD, so swaps performed on the OBD I '92-'93 Integra will require far less wiring. Earlier OBD 0 Integras will require additional modifications in order to function properly. Whichever model is being used, engine selection isn't going to vary much. It's recommended to select an OBD I engine, as that will require the least amount of wiring in most cases.

The most sought after of all Prelude engines, the H22A, is the most popular choice for DA/DB chassis swappers. The H22A from both the '93-'96 USDM Prelude Si-VTEC and the '92-'96 JDM Prelude Si-VTEC will both make excellent choices.

The H22A motors are among the least-expensive H-series engines but are extremely easy to locate in most junkyards across the United States. Non-VTEC H23A1 engines may be found in USDM '92-'96 Preludes Si's. The European and Japanese markets offer these same non-VTEC engines in both the '92-'96 Accord 2.3i and the '92-'93 Ascot Innova 2.3 Si-Z.

More expensive and somewhat rare H engines include the H22A found in the JDM '94-'97 Accord SiR. Many more-powerful Prelude variants exist in the '97-'01 model lineup, including the S Spec and the Type R. Unfortunately, these engines are OBD II and come with the unusable ATTS transmission. They should be avoided unless you're pre-

This Prelude VTEC engine is ready to be installed with a standard hydraulic-style transmission.

pared for some fairly extensive electrical work, as well as locating a different transmission.

Speaking of transmissions, with the exception of the ATTS units, several Prelude and Accord transmissions are totally compatible with both the second-generation Integra chassis and Place Racing's bolt-in kit. Any of the H22A transmissions are the best choice. The H22A transmissions are best suited for high-performance applications because of their being adorned with the closest gear ratios of all of the compatible units. If you're serious, check out the optional JDM H22A transmission with the factory-installed limited-slip differential. This is by far the best pick for this chassis.

Other compatible transmissions include all '92–'96 Prelude and '90–'97 Accord units. Although not exactly designed to work in conjunction with VTEC engines, these transmissions will provide you with exceptional gas mileage thanks to their much longer gears.

ECUs and Wiring for the H Series

The H-series engines were only offered in OBD, so '90–'91 Integra owners will be forced to modify their vehicle's underdash wiring harness in order to accept the new ECU. Since plug-in adapter harnesses aren't readily available for this conversion, adding proper plugs will need to be done manually. Once you get the right plugs connected, the addition of an OBD I ECU is simply a plug-in process. OBD II con-

versions may be performed in much the same way, providing the proper ECU plugs are acquired from an OBD II vehicle. Remember, '92–'93 Integras only require a conversion if an OBD II ECU is desired. OBD I computers of choice include the P13 Prelude VTEC ECU for all VTEC swaps and the P14 Prelude ECU for all non-VTEC swaps. OBD I versions may be acquired from any '92–'95 USDM or JDM Preludes with the corresponding engines. OBD II ECUs may be acquired only from the '96–'01 JDM models. Avoid the USDM

OBD II computers, as they're either equipped with the anti-theft immobilizer system, which renders them useless for swappers, or a mess of additional emissions sensors. Other compatible VTEC computers include the OBD I P72 Integra GSR, the P30 Del Sol DOHC VTEC, and the modified P28 Civic EX and Si ECUs. All of these boxes will work but are usually only used if the P13 computer cannot be located.

If you're using an OBD I engine and the proper modifications have been made to the vehicle harness, the original Integra engine harness can now be modified. Begin by swapping out the fuel-injector plugs for Prelude connectors. This allows you to attach the wiring harness onto the fuel injectors, giving a solid starting point.

With that in place, you can determine which wires need to be lengthened and which connectors need to be swapped out. First and foremost, VTEC engines will require the addition of wires for the VTEC pressure switch and the VTEC solenoid. The pressure switch can be skipped if a JDM VTEC engine is used without a pressure switch provision, providing the corresponding ECU is being used. For those folks swapping into the '92–'93 GSR, VTEC is already

By placing the original engine harness on the donor engine, you'll be able to determine which plugs and wires need to be cut, lengthened, or swapped. Begin by attaching it to the fuel injectors.

wired into the engine harness and needs only connector swapping and wire lengthening.

Aside from VTEC wiring, when using the P13 or P14 ECUs, the intake air bypass must be wired into place to allow you to use the secondary intake system. You must accommodate for an EGR valve as well when using the Prelude computers. When using any of the other ECUs mentioned above, this step can be skipped, since these computers won't recognize it anyway.

Further modifications include grafting the H-series alternator wiring assembly onto the Integra harness with the plastic cover and all. This will allow for a clean fit when wrapping it over the engine's valve cover. You'll also need to modify the harness to include the oil-pressure sensor plug, the reverse-light connector on the transmission, and the distributor plugs. Next, you need to lengthen the wiring for the idle air control valve and the oxygen sensor. A provision for a knock sensor must also be accounted for in the wiring harness for all engines.

Lastly, for those with '92–'93 OBD I Integras, adding an injector resistor box will be necessary for the Prelude injectors to function with the Prelude ECU. Failure to wire up one of these boxes will result in burnt drivers inside the computer. Additionally, on H engines equipped with external distributor coils, you'll have the option of using the current setup or converting back to the internal system by obtaining the proper coil.

Installing the H Series

Once the modified Integra wiring harness is adapted to the engine, the Place Racing engine mounts can be installed. Start by removing the Integra's rear engine mount and replace it with Place Racing's version. You'll find that it bolts easily into place on the rear crossmember. The driver-side mount can be attached to the original pocket mount on the frame rail, while the corresponding bracket on the engine will be that of the H series.

Slip a '92–'95 manual-transmission Prelude rear engine bracket into position while lowering the engine into the bay with the tranny side at a downward angle. Attach the bracket to the rear mount, not to the engine, and then lower the engine and transmission into place. The passenger-side bracket can now be bolted to the frame, allowing the top transmission mount to attach to it.

This Place Racing driver-side engine mount allows this whole procedure to be a bolt-in affair. Just think: this job used to require welding.

When the side mounts are fastened down, the rear bracket can be attached to the engine and tightened. Then fasten down all three mounts. Once the engine is installed with the Place Racing kit, you'll find that it leaves plenty of clearance under the hood and below the vehicle. There is actually a little bit more room under the hood of this generation of Integra in comparison to the '94–'01 chassis after an H-series swap.

Axles and Suspension Made Easy

Once the engine is fully installed, the underside of the vehicle can be finished off starting with the axles and the suspension. Since you'll need custom axles for this swap, you can reuse all of the original Integra suspension components. Reattach the lower control arm to the shock fork, and put the suspension back together in the same way it was removed. Unlike the H-series swap into the '88–'91 Civic chassis, the factory radius rods can be reused.

You'll need a custom front crossmember from Place Racing because of concerning clearance issues. The Place Racing unit will bolt onto the factory locations and has provisions for the radiator and radius rods. The rods may be reattached to the new crossmember in the same way that they were removed from the old unit, using all of the original washers and bushings. Custom axles can either be measured and fabricated, or purchased from Place Racing as premade pieces. Custom axles are mandatory due to differences in the inner diameter of the wheel hubs between Integras and Preludes. Once acquired,

When dealing with OBD I Integras, a resistor box such as this one will need to be added. Either bolt it in place or zip tie it somewhere safe once it's wired up.

The transmission mount is just as easy as the driver side. Installing the engine is just another bolt-on process that takes no time at all.

This Place Racing crossmember is designed for use with the H-series engine. You'll find that it easily bolts onto the factory locations.

these will slide right into place just as any other axle would.

Shifting, Prelude Style

Since the Prelude transmissions use a cable-operated shifter, the Integra's standard rod and lever unit must be tossed. Unfortunately, this is far from a straight bolt-in procedure.

As you'd expect, there are a couple of ways to do it. When you buy your swap kit from Place Racing, you might want to pick up one of their shifter housings too. This box-type device bolts up to the underside of the chassis allow-

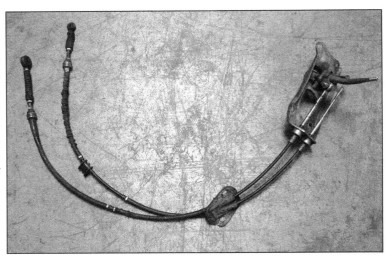

Shifter cables and assemblies like these from the '90–'97 Accord work great. This was purchased as an entire assembly to avoid a mismatch of components.

Early Style Auto-to-Manual Conversions

Throughout this book, it's generally assumed that a vehicle with a manual transmission is being used and that an engine with a manual transmission is being installed. For some folks though, this is not always the case. Many people just might consider taking on a project that involves an automatic vehicle and converting it to a manual. Others may have an automatic vehicle and wish to retain its slush-box auto trans. Either way, if you're using a vehicle that would have a cable-style transmission, were it originally a five speed, you can usually handle both of these situations.

If a manual conversion is to take place, then a plethora of associated parts will be necessary. You'll need a pedal assembly, clutch cable, and passenger-side transmission mount, just to name a few. Of course, you'll need other manual conversion parts also, and these will be mentioned throughout the book depending on which swap you're doing. Holes will need to be drilled and work will be required underneath the dash, not to mention quite a bit of additional wiring.

If you decide to go the second route and keep your automatic, it will prove just as time consuming. Most engine-mount companies manufacture their bolt-in kits for use only with manuals. If you decide to stay automatic, then you'll need to fabricate your own custom brackets. By the time all is said and done however, using an automatic in a vehicle with an engine swap rarely makes any sense.

The manual conversion won't be a straight bolt-in procedure if you're dealing with a couple of the vehicles that use the cable-style transmissions. Since the pocket

Watch out for the shorter automatic torque converter bolts. These are not long enough to mount a flywheel securely in place. Pictured from left to right: clutch pressure-plate bolt, flywheel bolt, torque-converter bolt.

mounts differ from the '88–'91 Civic to the '90–'93 Integra, they must be removed from the chassis and replaced with the manual version. Pocket mounts are the sheetmetal brackets that are spot welded to the frame or chassis. They house the engine mounts. If a job like this appears to be too much, you might consider Place Racing's bolt-in mount, which adapts the B-series manual transmission into these vehicles. All other engine swaps into these chassis will require welding when concerning the transmission. The '84–'87 Civic and '86–'89 Integras can be easily converted with the mount kit of choice. The '86–'89 Accord and '88–'91 Prelude aren't much more difficult. They will simply need to have their right side pocket mounts swapped for manual versions.

To wrap things up, a few electrical issues will need to be taken care of. The two reverse-light wires will have to be moved from their position on the shifter assembly to the sensor on the transmission in order for

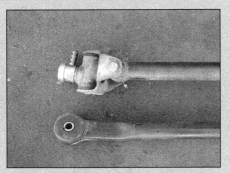

Shift rods of some sort will be necessary. You can usually find the ones you need in a salvage yard, or at your local Honda dealership's parts department. The rod on top is the gearshift rod, and the bottom is an extension rod.

When converting your auto engine to manual, you'll need to remove the flex plate and find yourself a compatible clutch.

the back-up lights to work. Next, you'll need to remove the shift-lock solenoid from the ignition. This will allow you to take the key out of the ignition once it's turned off.

This adapter box from Place Racing allows you to eliminate the Prelude or Accord shifter plate. Bolting this to the underside of the car hides the cables away.

ing the cables from any '90–'97 Accord or any '92–'01 Prelude to fit.

Once attached to this special box, the cables can be routed underneath the vehicle, over the rear engine bracket, and finally attach to the gearbox. For those who prefer to go at it themselves, the cables can be installed without the special adapter box. Either way, an opening large enough for the Prelude or Accord shifter assembly to sit flat is going to need to be cut. If you're not using the box from Place Racing, you'll want to position the stock shifter plate above the original shifter opening and fasten it down with four 8-mm bolts and nylock nuts. When going this route, the cut must be made very precisely to allow proper clearance for the shifter to function. In addition, keep enough material left over to which the plate will be fastened. Once attached, the cables can be run inside the vehicle and fished down a 2-inch hole cut about 18 inches in front of the shifter. Regardless of which method you choose, it's important that cables and shifter assemblies come from the same type of vehicle. Although many appear similar in size and shape, some aren't compatible with one another. Once finagled over the rear mount, the cables will attach to the transmission just like stock.

From Cable to Hydro

Since the Integra had a cable-operated transmission originally, you need

the Place Racing cable-to-hydraulic adapter kit. This will allow you to use the hydraulic-style Prelude or Accord transmissions. Without a kit like this, the vehicle would need to be retrofitted with a new pedal assembly, a clutch fluid reservoir, master cylinder, and a hydraulic clutch-line system. With the kit, the original clutch cable is retained and works in conjunction with a slave cylinder that is attached to the bracket. Trust me; this piece saves hours of labor and quite a few bucks in parts, too. With the clutch system underway, the throttle cable bracket from Place Racing will allow the original Integra throttle cable to be reused. Using this bracket will eliminate the need to crawl underneath the dash, unclip the original cable, and install a new one.

Fuel Rails, Lines, and More

As with most any Prelude engine transplant, the fuel system is going to do anything but reattach with ease. Due to the differences in the fuel injection feed line and fuel rail, a compromise will have to be made. The easiest thing to do is swap out the feed line altogether for a Prelude unit with the banjo hose ends. For those rare Prelude fuel rails that feature the inlet port on the opposite side of the engine, a swap will be in order. Rather than making a dangerous fuel-line extension reaching across the engine bay, a better solution is to replace the fuel rail altogether. Using

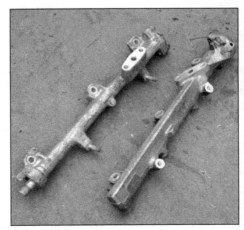

Here's a comparison of the two different types of Prelude fuel rails. The unit with the fuel feed port on the passenger side (on the left) works best on the Integra.

the '92–'96 USDM Prelude fuel rail will make quick work of this problem. Be sure to use new aluminum crush washers when hooking things back up for a leak free seal.

Engine Cooling Guidelines

The original Integra radiator can no longer be retained when using the Place Racing front crossmember. They purposely ruled out the wider Integra radiator, because they foresaw that it would rub against the Prelude's exhaust manifold anyway. A radiator from any '92–'00 Civic will bolt right into place with the help of Place Racing's special adapter bracket. When shopping for a '92–'00 Civic radiator, it's important to note that the '94–'97 Del Sol DOHC VTEC and '99–'00 Civic Si units are considerably thicker, thus providing far superior cooling. These performance-minded radiators also have larger inlets and outlets, making them the perfect size for attaching Prelude or Integra cooling hoses.

You'll also need a slimmer radiator fan because the original piece will hit the slave cylinder. Mount an aftermarket fan on the front side of the radiator, but make sure it's a push-pull design. Mounting the fan inside the bumper area will provide much more clearance underneath the hood. Just make sure that you connect the wires so that it

This aftermarket Koyo aluminum radiator and Hayden fan will keep the new H engine plenty cool. If choosing a Honda unit, avoid any radiators that are not twin core in design.

spins in the proper direction. Install any '92–'96 Prelude upper and lower radiator hoses to finish off the radiator.

Two additional hoses will need to be connected to complete the cooling system, starting with the water inlet hose that exits the firewall. This hose can be reused and slid onto the engine's cylinder-head water pipe, along with the original clamp. The final hose needs to connect from the new engine's rear main water pipe outlet to the water valve on your firewall. Cut off a long section of heater hose to fit and finish off the install with an adjustable-tension hose clamp on each end.

A/C and Power Steering Walkthrough

Now that the engine is cool, if you want to keep yourself cool too, you're not entirely out of luck. Although it's not an easy process, the air conditioning can be retained providing you have the proper components. The same can even be said for the power steering. We'll cover that in a moment. In order to keep the A/C, you need to use the Prelude compressor bracket and compressor. Now all you need is to have custom lines fabricated for the compressor's inlet and outlet ports.

You can keep your power steering by reusing the Prelude engine's brackets and the corresponding pump for that engine. Although the Integra's power-steering feed line isn't compatible with the Prelude pump, the fitting on the end of the line can be cut off and replaced with one from the Prelude. Simply cut off the end section of the Integra line and braze or TIG weld the Prelude portion into place. Since this line sees extremely high pressures, you may want to either have it welded professionally or have a line custom made along with your A/C lines.

The H-Series in Summary

Thanks to the fact that all second-generation Integras are equipped with four-wheel disc brakes, safety issues related to braking will be very minimal. Moreover, since the H-series engines only weigh 85 pounds more than the B series, you'll be hard pressed to notice any adverse suspension effects either. Since the H series sits so well under the

hood of this Integra, retains the power steering and A/C systems, and provides an increase in brute power, it is definitely worth looking into. Apart from the fact that the heavier, larger engine and the harder polyurethane mounts are a bit unforgiving at times, this is one of the more reliable Prelude engine transplants

If you have a custom feed line fabricated, the original Prelude power steering components from the donor engine can be installed for a perfect fit.

that you'll encounter. Before the introduction of the bolt-in mount kit, a swap like this would have been as difficult as putting the H series into an '88–'91 Civic. You can thank the folks at Place Racing for making this Prelude-to-second–generation-Integra relatively easy and reliable for not too much dough.

Surprisingly, this is a very nice fit. Thanks to the bolt-in kit from the engineers at Place Racing, second-generation Prelude owners can now have H-series power without too much difficulty.

1992 TO 1995 CIVIC AND 1994 TO 2001 INTEGRA

Since the '92–'95 Civic, '93–'97 Del Sol (also a Civic), and the '94–'01 Integra are all so similar, it's sensible to group all of these vehicles together when talking about engine transplants. Of course, the Civic was introduced first, and with this new, more radical body style, Civic sales went through the roof. And when its bigger brother, the Integra, was unveiled two years later, it was perhaps one of the most definitive moments in Honda engine-swap history. It was with the introduction of the third-generation Integra that swap enthusiasts realized the uncanny similarities between the two vehicles. Once stripped of their body parts, both the Civic and the Integra share virtually identical engine bays. Of course, identical engine bays yielded identical engine mounting points. With identical engine mounting points, thoughts of interchanging Civic and Integra engines turned into actions. Although most enthusiasts know by now that the Integra engines are certainly compatible with the fifth-generation Civic, it should also be understood that any engine swap attempted in one chassis would be essentially the same in the other.

Civic Offerings

The EG/EJ/EH chassis Civic (usually all referred to as EG) first introduced in '92 was available in several different trim levels ranging from the EX, Si, LX, DX, VX, and CX. Civics of this period could be purchased in four-door, two-door, hatchback, or the sporty Del Sol body style.

Not all of these trim levels were available on every body style, and even though the EX and Si models were regarded as top-of-line, the latter was only available with the hatchback. These two were the most popular models for early buyers, as they were equipped with such amenities as standard A/C, power steering, and stereo systems. For the racing crowd however, the popularity of the lightweight CX and VX models remains unmatched. These

The crew at Honda Fiend works together to make short work of this B18C GSR swap into a fifth-generation Civic Si.

Stripped of its body panels, this Civic looks somewhat similar to the third-generation Integra. This is one reason why the Integra engine swaps into these '92–'95 Civics are so basic.

This Integra GSR is very similar in every way, shape and form to the Civic and Del Sol. However, since it's already equipped with a B18C1, an engine swap would be ridiculous.

Other than some pretty extreme body differences, the Del Sol is identical to the fifth-generation Honda Civic.

stripped-down Civic hatchbacks offer major advantages for those concerned about having the lightest and fastest car possible. Remember that subtracting weight is just like adding horsepower, and a vehicle that weighs well over 200 pounds less than its siblings will have a significant advantage on the racetrack.

Available engines on these entry-level Hondas include the most popular and powerful of the SOHC engines, the D16Z6. Producing 125 horsepower, these engines can be found only in the EX and Si model Civics and Si Del Sols.

Lowlier engines include the D15B7 in the LX and DX, the D15Z1 or D16A6 in the VX, and the CX's D15B8. The top-of-the-line '94–'97 Del Sol VTEC is not in need of an engine since it comes

with the same engine that many of these Civic owners plan on swapping in. Removing the 160-horsepower B16A3 would be a poor choice.

Integra Offerings

Integras produced during this period offered far fewer choices compared to its sibling. The RS, LS, and GSR models were the only three choices before the introduction of the Type R in 1997. The RS and LS models came equipped with B18B1 engines similar to those of the second-generation Integras. The B18B1 engines began life as OBD I and switched to OBD II in 1996. GSR models were equipped

with the more-powerful B18C1 engine, which just so happens to be one of the most popular donor engines for Civics today.

Lastly, the B18C5 Integra Type R engine is undoubtedly the most-coveted B-series engine of its time. With 190 horsepower in stock form, it goes without saying that these Type R models aren't in need of any transplant. In fact, the only models that should be considered for an engine swap are the 140-horsepower RS and LS. At the very least a B16A VTEC engine would be an excellent upgrade. Since all the DC-chassis Integras are anything but lightweight, they aren't exactly the most common vehicles to receive an engine swap. But the sporty interior and body

The D16Z6 is one of available several factory engines for this Civic. Out of all of the available SOHC Honda engines, this is one of the more powerful and most upgradeable.

Integra GSR

One car that's in no need of an engine swap is the '94–'01 Integra GSR. It is equipped right from the Honda plant with a 1.8-liter 170-horsepower DOHC VTEC power plant. Not too many engines available that will outperform this one. The GSR power plant is among the most common engines to be swapped into the Civic chassis, so Hondaphiles are well aware of what lurks under the hood. Although the GSR is trimmed with leather interior, larger disc brakes, and factory alloy wheels, it's obvious that performance enthusiasts are after the engine.

Integra GSRs such as these can pull mid- to high-14-second time slips at the drag strip with little modification.

lines of the Integra remain unmatched by the Civic, which often more than makes up for its tipping the scales at around 2,700 pounds.

Available Upgrades

Since all Civics and Integras share similar suspension and braking components, fifth-generation Civic owners need not look any further than the '94–'01 Integra for enhancements. The front knuckles and disc brakes off the GSR and Type R Integras are interchangeable with the Civic, so they make perfect upgrades when additional braking is deemed necessary. Rear disc-brake conversions are also possible with parts from any of the late-model Integras. Since these swaps are straight bolt-in procedures, Civics in need of superior braking have it made. Civic Si owners can overlook the rear brake upgrade, as their vehicles currently use a rear disc setup.

Suspension modifications can further add to a successful engine swap. Due to additional understeer from some of the heavier engine swaps, specially designed swap springs from Progress Suspension may be in order. Progress designed these springs with a spring rate to handle the additional weight of the B- or H-series engines and to help counteract any negative suspension effects.

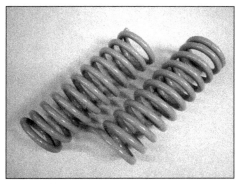

If you install a heavier engine, often you need to upgrade the suspension along with it. These Progress Suspension lowering springs were designed considering the extra weight. They'll help avoid any negative handling effects caused by the extra weight.

OBD I Vehicles

Although the Civics of this period maintain OBD I status, the Del Sols and the Integras switch from OBD I and OBD II. When dealing with any of the OBD I vehicles, it would be in your best interest in terms of electrical work to keep it that way. OBD I engines and computers can be found in the form of virtually every B series and H series available. However, OBD II conversions are sometimes necessary for the rare Type R transplants. If this is the case, a lengthy ECU electrical plug conversion will be necessary underneath the dash.

On the '92–'95 models, the computer will need to be located behind the passenger-side kick panel on the floor. The original vehicle harness plugs must be cut off and removed. The same procedure must be performed on a junkyard vehicle with the corresponding OBD plugs of the computer to be used. Once acquired, consult the service manuals for the vehicle and computer to make the necessary connections. After the plugs are soldered onto the vehicle harness properly, the new ECU will plug in with ease. Due to differences in OBD II connectors, you'll want to identify the year of the ECU to be used before obtaining the new connectors.

When dealing with the plugs found on Civics and Integras, notice that the OBD II '96–'98 units are different from the OBD II '99–'01 units. Be sure to match up the proper connectors with the new ECU before soldering them into place. JDM OBD II ECUs should be used in all cases in order to avoid unnecessary labor relating to the USDM OBD II ECU's several additional emissions sensors and anti-theft immobilizer systems ('99–'01 versions only). Unless you're prepared for some intense labor, the JDM units are your best bet.

OBD II Vehicles

Things are rather easy regarding the '96–'01 vehicles. If you use an OBD II engine, you can use any USDM or JDM OBD II ECUs on these vehicles. However, you may have to remove the built-in anti-theft immobilizer system in the '99–'01 Integra USDM ECU before using it. The JDM units do not have this feature though, so of course they're preferred. When converting to OBD I status, it isn't necessary to manually perform the underdash wiring conversion. Several manufacturers including Place Racing and Hasport Performance produce adapter harnesses that simply plug into the vehicle harness and allow you to use an OBD I ECU. Due to the differences in OBD II plugs, be sure to let the manufacturer know what year your vehicle is before placing your order.

B-Series Engine Swap

The '92–'95 Honda Civic (EG chassis) is without a doubt the most versatile and inviting chassis for a B-series engine. This should come as no surprise, since it shares the same engine bay characteristics as the '94–'01 Integra. It really doesn't make any difference whether you prefer a sedan, coupe, hatchback, or Del Sol body.

With that being said, it only makes sense to mention B-series possibilities concerning the '94–'01 Integra chassis here as well. You might be wondering why the DC-chassis Integra is even mentioned as a swap candidate, since it is equipped from the factory with more B-series engines than any other Honda or Acura. This raises a good point. Even though almost every Integra has always had some sort of B series under its hood, upgrading to a different B-series engine can sometimes be enough of a challenge to be worth mentioning here. Let's remember, regardless of how similar two engines may be, if one is being changed out for another, it constitutes a swap.

If you look around at the next race or event, you might be surprised how many '94–'01 Integras actually do have different B-series engines underneath their hoods than they were originally sold with. For example, with the price of an Integra Type R as high as it is, a somewhat economical solution would be to find an inexpensive RS model with a tired engine and swap in the Type R drivetrain. This is just one of the many reasons that the third-generation Integras have become such viable B-series swap candidates. Let's also not forget that they're Integras, and it's not going to be too difficult to swap in another Integra style engine, no matter how different they may appear from one another.

All of the B-series engine transplants into '92–'95 Civic, '93–'97 Del Sol, and '94–'01 Integra chassis are pretty much straightforward and uncomplicated. It's fair to say that anyone who is able to remove the original Civic or Integra engine successfully would be competent enough to install any of the B-series engines back into its place. With the correct parts and the right tools, these transplants are downright basic. So basic, in fact, that professionals have been known to do complete swaps of them in a matter of hours. Even the inexperienced can complete this project in a mere weekend.

B-Series Engines and Transmissions

Among the most common B engines to be transplanted into any Honda or Acura is the ever-so-abundant B16A. With so many different versions and years of production, and so many available at the wrecking yards, finding the right B16A can get just plain confusing. Let's briefly cover some of the applicable B16A engines for this particular chassis, which are only those that are equipped with the hydraulic-style transmissions.

We'll begin by mentioning the B16A3 from the '94–'97 USDM Del Sol DOHC VTEC. With a chassis similar in many ways to the DC Integra, the Del Sol will provide the installer with almost everything necessary for a complete changeover. Similar to the B16A3, its JDM counterpart B16A can be acquired from the '92–'97 JDM CRX del sol SiR, as well as from the '92–'95 JDM Civic SiR and SiRII. Most B16A engines are nearly identical to the USDM Del Sol DOHC VTEC power plant, and will usually be sold with most of the parts necessary for the transplant.

Available OBD II engines include the B16A2 located underneath the hood of the '99–'00 USDM Civic Si, and the JDM counterpart B16A, which can be found in the '96–'00 JDM Civic SiRII engine bay. Rounding off the list of B16 engines is the ever-so-rare B16B Type R engine. Although not exactly a B16A, the B16B found on the '98–'01 JDM Civic Type R can attribute most of its characteristics and its heritage from the B16A family. With over 185 horsepower from only 1.6 liters, this is one of many reasons why one would want to swap a B-series engine into a DC Integra. Of course, this would be a perfect upgrade to the EG chassis as well.

A late-model GSR engine is among the most popular donor engines for these vehicles. The USDM transmission, wiring harness, and mounts are all compatible with these cars.

In addition to all these B16A engines, you'll also find an equal amount of B18 engines that are equally installer-friendly for this swap. Of course, most Integra owners should be aware of this already, since these Integras already have a B18 of some sort under the hood. Nevertheless, here is the rundown of compatible B18 drivetrains for the '92–'95 Civic, '94–'01 Integra, and Del Sols as well.

Sticking with engines that use the hydraulic-style transmission, most everyone should be already aware of the original B18B1 in all '94–'01 RS and LS Integras. The upgraded engine, otherwise known as the B18C1, can be found in any '94–'01 Integra GSR.

Similar engines referred to simply as the B18C can be yanked from the '95–'97 JDM Integra SiR-G. Available Integra Type R engines include the '97–'01 USDM B18C5 and the JDM B18C. These are easily distinguished from the GSR B18C by several characteristics. The Type R version will be clad with a red colored valve cover. Other R-engine indications include the tubular racing header on the JDM version, or the racing-inspired intake manifold that looks more like one from a B16A than a GSR. Perhaps the most clear indication of a true Type R engine can be found in the cylinder head. The GSR cylinder head is labeled P72, whereas the Integra Type R is more akin to that of a Del Sol VTEC.

Other notable B-series engines worth swapping include those from the

Engine ID Codes

One thing you'll notice when inspecting the engine identification code on a JDM engine is the lack of a key reference number. While USDM and EDM engines are differentiated by a numerical digit (following the A in B16A for example), their JDM counterparts are not. When differentiating engines here in the United States, matters are quite simple. The B18C1 and B18C5 are clearly distinguished by the 1 and the 5, so it's easy to tell the two apart. This is the case with all USDM engines. Unfortunately, in JDM land, these numbers are nonexistent. The only way that you'll be able to tell a Type R from a mere GSR is by using your own keen engine-identifying skills.

Here are the ID numbers on a JDM engine. This can be identified as a JDM engine by the fact that it doesn't have any digits following the B16B.

The 1 digit following the B18A is a quick reference when trying to find the country of origin of a particular engine. This is a USDM Integra engine.

CRV family. With 2.0-liters of displacement, these engines respond very well to the addition of a VTEC cylinder head. B20s are becoming increasingly more popular than the LS/VTEC conversion, and the added displacement makes this conversion all the more worthwhile. Possible CRV donor engines include the B20B found in the '97–'00 USDM CRV and the more powerful B20Z found in the '99–'00 JDM S-MX. Both engines are similar in external size and shape, while their major differences are found internally. A word of caution: Stay away from any of the B-series engines found in the '88–'91 Preludes. The Prelude B20A3, B20A5, and B21A1 engines aren't compatible with this Civic and Integra chassis and shouldn't be confused with the more up-to-date B20s found in the CRVs and SMXs.

This late-model Integra hydraulic transmission can be identified by the style of its input shaft. These hydro units also feature studs on the top to which the top mount is attached.

Now that all the B-series engines are covered, it's time look at your gearbox options. Unfortunately, things in that department are just a little more confusing than selecting the engine. With every single Integra and Civic B-series transmission able to properly function in this particular chassis, we have quite a bit of ground to cover.

As with the engines, we'll only be covering the hydraulic-style transmissions for a couple of reasons. Although the introduction of the adapter kits from Place Racing, Hasport Performance, and HCP Engineering allow the use of cable-operated transmissions into the later-model chassis, these still aren't the

Although the sticker can identify many transmissions, these often fall off, or worse yet, are switched. Check the stamping on the block; there's no confusion here.

most popular of choices when dealing with the newer vehicles. There are still almost a dozen compatible hydraulic-style transmissions.

Interestingly enough, every single one of these gearboxes is identical in size, shape, and overall appearance. The good news is that that means that they're all interchangeable and compatible with all B engines. The bad news is that without taking them apart for inspection, the only way to differentiate an expensive Civic Type R limited-slip equipped transmission from a cheap USDM Integra RS unit is a sticker. With that knowledge, it should be understood that when selecting a transmission that is separated from the engine, it would be wise to make your purchase from a reputable junkyard that offers a warranty or exchange policy. The scary thing is that these do get mixed up from time to time, and without the proper identification sticker, who knows what you'll be getting.

In addition to all of the B16A and B16B transmissions that are available with the engines listed above, all the late-model Integra transmissions will work too. These include both USDM and JDM Integra Type R transmissions, as well as all of the USDM and JDM '94–'01 Integra units. The available Integra transmissions alone make up six of the possible transmission choices. Driving style and gas-mileage expectations should also be taken into consideration before purchasing a transmission. All the VTEC units have significantly shorter

gear ratios as opposed to the RS and LS units, and so will be optimal at the racetrack, yet not as forgiving when it comes to gas mileage. Compromises are sometimes crucial, and sacrificing good gas mileage for speed is sometimes necessary.

ECUs and Wiring for the B Series

The DC Integra can be split up into two basic categories when dealing with the electrical system, thus making things simple. The original DC Integra and all of the EG Civics are equipped with the first OBD system and remained that way until the introduction of the 1996 models. Del Sols were also mainly OBD I, excluding the 1996–1997 model years. OBD II was implemented on these vehicles in 1996 and remained so until 2001. Although all of the engines referred to above will easily fit into any of these OBD I vehicles, the proper engine selection will ensure a minimal amount of wiring labor. It goes without saying then that if you want to keep the swap simple, get an engine and computer with the same OBD system.

When dealing with the '92–'95 OBD I vehicles, swapping in an OBD I engine and ECU will take the least amount of work. These ECUs will simply plug into the vehicle harness of the Civic or Integra.

This P28 computer is a popular alternative to the DOHC ECUs. Although its origin is from a SOHC vehicle, it can be reprogrammed for the B-series engines.

The same thing goes for '96–'97 Del Sols and '96–'01 Integras and OBD II engines and computers. In these instances, choose an OBD II computer that corresponds with the plug type of your vehicle (remember, there is more than one type of OBD II).

Jumping back and forth between OBD systems is certainly an option for those who aren't worried about a little wiring. In some cases, it's an option that is unavoidable. For instance, for those who wish to run almost any of the Type R engines, Civic or Integra style, will have to deal with mismatching OBD issues if working with a '92–'95 chassis. You'll need a wiring conversion for underneath the dash to accommodate the new ECU connectors. Be sure to use the JDM units, so you won't have to worry about emission sensors. If an extensive wiring conversion isn't desired, you can always choose to use an OBD I ECU on the newer engine, since it won't make much difference (the necessary steps are listed in the next paragraph). Keep in mind that even though all of the OBD I DOHC VTEC ECUs will work with the Type R engines, they won't provide you with the optimal fuel and timing maps. Note: A couple of rare JDM OBD I Type R ECUs exist, but obviously they're pretty hard to come by.

If you're planning an OBD II to OBD I conversion, you have a wide selection of compatible OBD I computers to choose from. Those wishing to go this route can use either an OBD I or II engine; it won't make much difference. You'll need to modify the distributor plugs and add an underdash adapter harness for the computer to complete the conversion for the OBD I ECU. Such harnesses are readily available from the likes of Place Racing or Hasport Performance, just to name a couple. Again, if using an OBD II ECU, be sure to get the proper year of computer that corresponds with the OBD II type of the vehicle. For OBD II equipped Del Sols and Integras, either '96–'98 USDM versions or '96–'01 JDM versions can be used. Avoid the '99–'01 USDM units with the anti-theft immobilizer system.

Staying with the vehicle's original OBD type will make the wiring portion downright simple. It will require the occasional lengthening of sensor wires and the addition of a VTEC pressure switch, VTEC solenoid, and knock sensor wiring for VTEC-equipped engines.

Additional wiring modifications include provisions for the evaporative purge solenoid on all B engines, and the intake air bypass valve for GSR engines.

You must add wiring to the harness for the VTEC pressure switch shown here. The VTEC solenoid wiring is to its right.

When switching OBD systems, besides the ECU wiring, distributor plugs will need to be modified because the early distributor has two electrical connectors, while the newer version has only one. Compare wiring schematics for each distributor, side by side, to ensure that this process takes no more than a few minutes.

As mentioned earlier, in certain cases additional sensors need to be wired to the computer for them to function properly. These modifications are of an intermediate level and can be accomplished quickly with the appropriate service manual comparisons. For Civic owners looking for the utmost in pain-free wiring, you might consider using one of the USDM Integra harnesses. Although this is more expensive than reusing the original harness, in many cases these Integra units will eliminate the majority of the wiring labor.'

Installing the B Series

Locating and installing the proper engine mounts for a stock-type fit couldn't get any easier. If you're swapping into an Integra, it's important that all of the original mounts are set aside and saved. These can be recycled later on the new engine. In many cases, brackets and mounts are even left on the donor engine from the junkyard. If they match up, compare them with your original mounts and install the ones that are the least deteriorated. In other cases, some brackets and mounts left on junkyard engines are incorrect and must be swapped for your original mounts. Per-

When installing the B20 engines, the post mount will have to be swapped out for the appropriate '94–'01 Integra unit. This bracket can be removed by taking off the plastic timing-belt covers.

fect examples include the '99–'00 Civic Si, '94–'97 Del Sol DOHC VTEC, and all CRV engines. Although these brackets are B series in nature, the vehicles they came out of required different mounts. Be sure to remove the rear engine bracket on the Si and the CRV, and the left-side engine bracket located behind the timing belt cover on all three, as these are all incompatible now with the Integra. These components may be replaced with the original Integra units.

The Civic and Del Sol chassis are a different story and will require a little bit more work since you can't reuse all their mounts. Depending on which Civic or Del Sol you have, the driver-side engine mount will differ slightly. All '94–'95 EX, Si, VX Civics, and all Del Sol Si models come equipped with an Integra-style left motor mount, so the Integra-style post mount may be used. For all other Civics and Del Sols, the left mount will be black in appearance, as opposed to the brushed aluminum Integra version.

These mounts will work in a different way, providing they're used with a '94–'97 Del Sol DOHC VTEC or '99–'00 Civic Si engine block post mount. Both of these driver-side engine mount combinations will provide you with a factory-type fit and identical engine placement. Either Civic or Integra mount combination may be used on either chassis. However, the same rear engine brackets that won't work on the Integra, won't work with the Civic either.

For the Civic to B-series transplant, you'll also need a '94–'01 Integra or

Once the correct post mount is installed, in certain cases you can reuse the original Civic or Integra engine mount on the driver side of the vehicle.

This front transmission bracket stabilizes the transmission during engine movement. Although some choose not to install it, it's highly recommended.

Certain Civics and Del Sols use this different left-side mount. You need to install a different post bracket on the block if you're going to use this mount.

'94–'97 Del Sol DOHC VTEC rear engine bracket. The '99–'00 Civic Si and CRV brackets won't work in this situation, as they're way too short. When connecting the rear bracket, the original rear rubber engine mounts can be reused on all of these vehicles, whether they started life with an automatic or manual transmission. The original top transmission bracket of any of the vehicles receiving the transplant can be reused as well.

Last, the front transmission bracket and air-conditioning bracket can both be sourced from a '94–'01 Integra, '99–'00 Civic Si, or '94–'97 Del Sol DOHC VTEC. A bracket will also be needed to attach the front left of the engine block to the frame rail. If A/C is to be retained,

the '94–'97 Del Sol DOHC VTEC engine bracket is your best option, as it will allow you to attach the Civic A/C compressor. If A/C isn't present, any of the late-model B-series brackets will attach the engine to the frame rail in the same manner. Be sure to reuse both of the original rubber stopper mounts underneath the frame on the Civic or Del Sol. These are vehicle specific and cannot be interchanged for Integra units.

When lowering the engine into place, the only bracket that should be fixed to the engine is the left-side piece. All of the rubber mounts should remain on the vehicle, but all other brackets can be set aside for now. With the transmission side dropped below the frame, attach the left-side engine bracket to the

Do not remove these special rubber engine vibration mounts from the vehicle. The Integra versions are not compatible with the Civic and vice versa.

left side motor mount. The rear engine bracket can be slipped into place at this time. Make sure the rear bracket is installed before raising the engine and attaching the right mount, because it will be difficult to finagle it into place afterwards. With the three mounts attached, the front two brackets can be installed and attached to the block and the original rubber stopper mounts underneath the frame. With all five mounts in place, you'll find that engine sits under the hood exactly like the original one did.

As with any transplant using the stock motor mounts, you can be assured of optimal engine placement, as well as adequate hood and ground clearance. The B-series swap into the Civic, Del Sol, or Integra is no exception. Cutting and welding won't be necessary for this simple bolt-in procedure because clearances between the engines and frames are exactly the same.

Axles, Linkages, and Cables

In keeping with the theme of simplicity, the axles and suspension can be reused. All late-model Integra axles are the same, as is the basic scheme of the suspension. The original axles and suspension components may be reused and installed once the engine is in place. As you'd expect, the shift linkage is no different than the suspension. Hang on to the original Integra unit, as it will work with any of these transmissions. If you're swapping into a Civic, Integra axles and shift linkage will complete the underside of the vehicle. Avoid the Del Sol DOHC VTEC linkage, as it won't fit into the Civic chassis.

Reattaching the clutch line is also straightforward on all vehicles. To select the proper throttle cable, you'll need to take into account which engine is being used. For instance, if using a GSR engine, be sure to use the appropriate GSR throttle cable. The lengths of individual cables are different, and for proper throttle response, it's mandatory to install the right one. If you'd rather install an adapter bracket and reuse the original throttle cable, give Place Racing a call.

You can attach the proper throttle cable to the original pedal assembly with this clip-on end. Don't forget to also attach it properly to the firewall with the supplied grommet.

Fuel and Cooling Affairs

As usual with these simple swaps, you just need to reattach the original fuel injection feed and return hoses to the appropriate locations. Be sure to replace the aluminum crash washers at this time. Some Civics use an incompatible fuel injection feed hose that must be replaced with an Integra-style unit. This may be expensive, but it's a good alternative to cutting and splicing the line with the hose end from the donor.

These Civic radiators will work, however, you should upgrade to the twin-core unit from the Del Sol DOHC VTEC. Some of these B-series engines will require the additional cooling.

Moving onto other fluid systems, note that the original radiator may be retained as well as both the cooling fan and A/C condenser fan. Upper radiator hoses may be selected by choosing one that corresponds to the new engine. The original lower radiator hose may be reused if it's still in good condition if you're swapping into an Integra. Since the Integra radiator is equipped with the larger inlet and outlet diameters, the original clamps may be reused with the new hoses. Civics will need to install adjustable-tension hose clamps in order to squeeze the larger hose onto the smaller water necks. Radiators from '94–'97 Del Sol DOHC VTECs and '99–'00 Civic Si's feature the larger water necks and will bolt right into place on these Civics. They provide a perfect fit, and have far superior cooling capabilities due to their twin-core design.

Maintaining A/C and Power Steering

To finish things up underneath the hood, reattach of the A/C and power steering systems. When dealing with the Integras, you'll find that the original A/C compressor will bolt right back onto the original bracket just as it was removed. All of its lines and components can be reinstalled in their original locations. Civics and Del Sols will use the Del Sol DOHC VTEC bracket mentioned earlier in order to use the Civic compressor and lines. In most cases, if you add the power steering brackets from the new engine, you can use either the Integra's original pump or the one from the donor engine. However, in certain situations there will be a conflict of parts.

Keeping your power steering is easy, providing you have the proper components. This Integra pump and line fits perfectly on the new engine.

With the '98 Integra came a new power steering pump. This means that attaching the original pump to the upper bracket that came with this newer donor engine won't work. In these rare cases, the correct upper bracket or pump combinations must be used.

A Finished Project

Except for the confusion with the power-steering system, you can see how easy installing a B-series engines into the EG and DC chassis is. The nice thing about this transplant is that when it's done, you won't need additional parts because of inferiorities that already existed in the vehicle. The suspension and brakes are a perfect example of this. All B-series engines weigh approximately the same, so suspension and braking won't be an issue on the Integra. Even the lowliest of DC Integras is equipped with four-wheel disc brakes and a sporty suspen-

In addition to the Integra brake conversion on the Civics, master cylinders and brake boosters may also be interchanged. By installing the larger diameter units, braking may be enhanced.

This B18C1 in this '92–'95 chassis is a perfect demonstration of how well B engines fit. USDM swaps like this one are totally emissions legal.

sion, so upgrades in these areas should be done by choice and not by necessity.

As for the Civic, you can juice up your brakes by simply swapping in front and rear Integra components. Compatibility between the two chassis makes a four-wheel disc brake conversion a bolt-on process. For further suspension enhancements on the Civic, Progress Suspension has specially designed engine-swap lowering springs that take into account the additional 100 pounds of the B engine.

The quick installation process and the ability to recycle many of the old parts contribute to keeping the overall costs of this transplant low. However, if you're in the market for any of the Type R engines or even the B18Cs, you'll soon find that these are some of the most expensive used Honda and Acura engines that you're likely to encounter. The average B-series engines are priced in the middle range in comparison to others on the market today.

The mid-range cost, abundance of horsepower, and stock-like feel and appearance are qualities that will make most enthusiasts satisfied with this swap. You can expect any of these transplants to perform reliably for years to come.

H-Series Engine Swap

For the sake of avoiding repetition, when discussing H-series engine transplants into the fifth-generation Civic, we're also talking about the Del Sol and the third-generation Integra. Since all three share virtually identical chassis, the transplant process will be identical.

This H22A1 came from a '93–'96 Prelude Si VTEC. It's one of the largest engines that will fit into these chassis; it also has the most horsepower.

When it comes to Prelude engine swaps, the Honda engine swap industry has definitely come a long way since the mid-1990s. The original Prelude engine swaps performed on the EG chassis Civics and DC chassis Integras suffered from broken engine mounts and snapped axles on a regular basis, among other problems. However, with plenty of research and a vast supply of new transplant parts, the H-series swap has become nothing short of reliable.

Today, there really isn't a more popular donor engine than the Prelude H22A. With 200 flywheel horsepower on tap from the JDM version, it stands as one of the most powerful four-cylinder production engines available. With more torque on demand than almost any other engine in this book, there'll be no turning back once you've been behind the wheel and felt the power of the H series.

As enticing as it seems, this additional power isn't going to come easy. A typical H-series swap into the EG or DC chassis would normally take a skilled mechanic approximately two to three days. With a minor amount of fabrication and some tricky wiring, novices might find themselves wrapped up in this for weeks on end. This is probably among the most challenging engine swaps discussed in this book. I'm not trying to scare anybody away from doing one of these; it's just best to know up front what to expect and how long it should take to complete.

H-Series Engines and Transmissions

Upon deciding on an H-series engine swap, it's important to see the different variations available. As with the B series, there are several different H-series engines and transmissions to choose from. In fact, more H-series engines will fit into these vehicles than you may think.

The H23A Prelude engine, although not VTEC, does boast the largest amount of displacement among all of the H series. At 2.3 liters, the H23A engine makes up for its lack of variable cam timing with some added torque. The most common place to find an H23A engine is under the hood of any '92–'96 USDM Prelude Si. Here in the United States, these engines are stamped H23A1

and are reasonably common in wrecking yards. A similar H23A may be found in the JDM Ascot Innova 2.3 Si-Z as well. The final choice for H23As is the H23A2, found in the '92–'96 EDM Accord 2.3i. Good luck finding one of those in the United States.

Now let's move on to the more popular VTEC H22A, which first appeared in the '92–'96 JDM Prelude Si-VTEC. These 200 horsepower donor engines are great for swapping, and best of all, they're relatively easy to find. The following year, Honda put this motor in the '93–'96 USDM Prelude Si-VTEC and labeled them the H22A1. With the major body-style change for the Prelude in 1997, the H22A4 and the ATTS transmission were added to all USDM Preludes with the SH badge.

Other vehicles that host the H22A include the '94–'97 JDM Accord SiR, the '97–'01 JDM Prelude SiR, and the often overlooked or unheard of H22A2 found in the '94–'96 EDM Prelude 2.2i. More powerful and extremely rare Prelude VTECs include the H22A in the '97–'01 JDM Prelude Spec S and the H22A7 of the '97–'01 EDM Accord Type R. Each of these engines are extremely hard to find and very expensive.

So there, you have it. That list of Prelude engines ought to keep you busy for a while if you're not sure which one to go with yet. To make things a bit more confusing, let's add some transmissions to the list as well. To avoid a horrendous amount of wiring for the additional computer box, you might want to stay away from the ATTS gearboxes. Although technologically advanced and extremely cool, the wiring involved is time consuming to say the least. According to folks who've dealt with these in the past, they really aren't worth it in the long run. Sticking with the standard Prelude VTEC gearbox isn't only easier, but it's downright cheap in comparison to the ATTS unit. Optional limited-slip differential versions of the JDM Prelude VTEC transmission are available as well, which gives you one more reason to stick with the easy route. H23A transmissions are compatible as well, but aren't recommended due to their longer gear ratios. Other compatible transmissions include the racing-geared units in the Type R and Type S Preludes and Accords. If you can get your hands on either of these, consider yourself extremely lucky.

That's not all. You might be surprised at one other group of transmissions that will work on this transplant. The other available choice is from the Accord. That's right, the '90–'97 Accord transmission will bolt right up to any of the H-series engines as if it was meant to be there. Although not exactly geared for race use, the Accord units are really cheap. This could be the perfect solution to your problems if cheaper costs and increased gas mileage are priorities.

ECUs and Wiring for the H Series

Since these particular H-series engines cross over into only two OBD zones, they make for a less-than-complicated electrical conversion in most cases. Since '92–'95 Preludes are equipped with the OBD I system, installing them into the '92–'95 chassis is an easy plug-in procedure concerning the ECU. When dealing with the '96–'01 vehicles, however, things get a little tricky. The '97–'01 USDM Prelude computers are rendered useless for swappers by their built-in anti-theft system. The anti-theft system is built into the radio and the ECU of the original vehicle, and most wrecking yards do not have the code. Therefore, for those of you who wish to use the USDM OBD II computer on your OBD II car, the '96 version is your only option. Due to the additional emissions sensors present on the USDM OBD II engines, you'll need to use a JDM ECU when converting a '92–'95 chassis to OBD II. All JDM OBD II ECUs are up for grabs and will work fine on both of these vehicles, providing you do the proper wiring.

There are really only three different situations that you might encounter with the wiring and electrical systems. For Civics and Integras that use the OBD I electronics system, it's easiest to select an OBD I engine. This situation will require the least amount of wiring and allow you to plug in the computer without a hitch. If you want an OBD II engine because of its lower mileage or increased horsepower, an OBD I ECU may be used here

as well. If an OBD II unit is deemed necessary, wiring modifications will need to be performed underneath the dashboard in order to accommodate the new ECU. Due to the differences in electrical connectors, the proper plugs found on an OBD II vehicle must be soldered into place, one by one, in order to accept the new computer. With the adapter completed, the OBD II ECU will function properly with the new engine.

The second situation involves the '96–'01 Integra and '96–'97 Del Sol with the OBD II system. Instead of converting a vehicle to OBD II status, most OBD II cars are converted to OBD I status. Due to their lack of emissions sensors and the ability to be reprogrammed, OBD I ECUs are usually a better route when horsepower is the main concern. For those folks with the OBD II vehicles, you need only purchase an adapter harness to run the OBD I ECU. Of course, if you're inclined, you can perform the ECU conversion on your own with the help of some schematics from Honda. Plug-in adapter harnesses have grown in popularity due to their ability to be removed from the vehicle so that it may be reverted to its original state. When using an OBD I ECU, it's important to note that the USDM OBD II Prelude distributors won't function properly and must be replaced with an older unit.

The last electrical situation for this swap involves an OBD II vehicle that the owner wishes to keep OBD II. In this case, either a '96 USDM ECU or a JDM ECU is mandatory. You'll need to convert the plugs underneath the dashboard on '99–'01 Integras, since they use a different style of OBD II plugs than the Prelude or the earlier OBD II Integra. Unless a Type S or Type R ECU is to be used, retaining OBD II should never need to be addressed and isn't recommended.

With the OBD situation straightened out, the basic wiring modifications that need to be performed to the Civic, Del Sol, or Integra engine harness will remain the same regardless of which OBD route is chosen. However, it's important to know which computer you're using before making any wiring modifications.

For those going the non-VTEC route, your most available option for OBD I swaps is the '92–'95 USDM Prelude Si ECU, known as the P14. For those interested in VTEC swaps, you have a couple more options. Although many DOHC VTEC ECUs will work, the two most popular choices include the OBD I GSR ECU and the OBD I Prelude VTEC ECU. These are more commonly referred to as the P72 and P13 respectively. The P72 computer eliminates the need to wire up the Prelude engine's EGR valve and intake air bypass valve. The EGR valve and the IAB will only be functional if you use the P13 computer.

You must use this vacuum box when installing the IAB system for a Prelude engine. Be sure to find one that doesn't have snapped-off vacuum ports like this one (they usually do).

As with all Prelude transplants, the original injector plugs will need to be cut off and swapped for Prelude connectors. An injector resistor box must be mounted to the chassis and wired into place in order for the saturated Prelude injectors to function properly. The purpose of the resistor box is to reduce the amount of electrical load on the drivers inside the ECU. Failure to wire this baby up will eventually result in a burned-up computer.

Since the alternator is located on the front of the H engine as opposed to the rear on the Civic and Integra, the alternator wiring on the Integra will need to be modified accordingly. Cut off the section of wiring on the Prelude engine harness that goes to the alternator and graft it on to the Civic or Inte-

By retrofitting the Prelude alternator wiring onto the Civic or Integra harness, the wires may wrap over the valve cover and down to the alternator. We added some tape and loom to this portion to clean it up.

gra engine harness. The added benefit of using the alternator wires from the Prelude harness is the way that they'll fit on the new engine. The Prelude's alternator wires normally wrap themselves over the valve cover in a pre-fit plastic piece. If you can salvage the wires and this plastic piece off the old Prelude engine harness, they'll help the wiring look as clean as possible.

Additional wiring modifications include lengthening the wires associated with the idle air control valve and the oxygen sensor several inches and swapping plugs on the transmission for the reverse light connector to plug in. Civic or Integra plugs are different from Prelude plugs. For those swapping into a Civic CX or VX, the wiring harness will need to be modified to accept a four-wire oxygen sensor as opposed to the original seven-wire unit. Final wiring modifications need to be preformed to the distributor if you're using a USDM engine. The USDM H-series engines are equipped with an external distributor coil that must be wired up accordingly. For those swapping in VTEC engines, the VTEC pressure switch, as well as the VTEC solenoid must be added into the harness. If your vehicle is pre-wired for VTEC, however, go ahead and skip these last steps. Interestingly enough, all H series engines, including the non-VTEC models, will need to have a knock sensor wired into place.

Installing the H Series

The process of bolting in any H-series engine has been made easier thanks to the folks at HCP Engineering, Hasport Performance, and Place Racing. Each of these companies manufactures engine mount kits designed for installing H-series drivetrains into the EG and DC chassis. Even when you purchase one of these mount kits, you'll want to make sure you have some stock Prelude components as well. You'll need the left-side engine bracket, which is bolted to the block, and a rear engine bracket, both found on any '92–'96 Prelude.

Before installing the engine, the Civic or Integra rear engine mount must be removed and swapped for one of the new aftermarket units. Some kits require you to completely remove the right-side engine bracket, while others just involve minor trimming and drilling a new bolt hole.

This Prelude post mount is used with all of the engine mount kits available for these transplants. You'll never need to swap this out for something else.

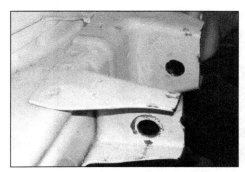

When using the HCP Engineering kit, you must partially cut off this passenger-side engine bracket. You'll need to drill another hole about an inch to the left of the original one that had to be cut off.

If one of your mounts needs to be removed, you'll need to drill out each of the spot welds individually for it to come off nicely. Place Racing provides a special drill bit with its kit for properly drilling out the spot welds. Check with the instructions supplied with your kit before cutting anything.

At this point, depending on which kit you use, the right side mount can either be attached to the top of the transmission in the original bolt holes or bolted to the underside of the frame rail. This all depends on which motor mount kit you use. With the left side bracket in place and the aftermarket mount in position on the driver-side frame rail, the engine can be lowered down. As with any Honda engine, lower it so that the transmission ends up below the passenger-side frame rail. This can be achieved by using an adjustable load-positioning bar on the end of your engine hoist. After attaching the left mount, the left side of the engine can be raised back up to attach the right mount. The manual transmission rear engine bracket can be finagled into place, sliding onto the rear mount with the help of some lubricant. If you aren't using a crossmember or front mount, then the three mounts can be tightened up at this time.

If you spend the money on a decent engine mount kit, you'll find that the Prelude engine fits under the hood better than you would have thought possible. Although the engine doesn't actually hit the chassis in any spots, there are a couple of locations close enough to mention. The closest spot the engine finds itself to the frame is located on the driver-side shock tower.

Jon Spackman of Holeshot Racing makes good use of his air tools when tightening the engine mounts down.

Depending on the kit, the valve cover sits within a few millimeters of the chassis. Luckily, the engine does not rock in this direction. Common solutions to this closeness include trimming the valve cover or slightly denting the shock tower with a ball peen hammer. Many folks cut their valve covers to expose their adjustable camshaft gears anyway, so this has always seemed to be the best solution for added clearance. Other clearance issues arise between the intake manifold and the firewall. With just a fraction of an inch between the manifold and the firewall-mounted brake lines, a broken rear mount could result in your vehicle being unable to stop. This should be reason enough to invest in a good set of mounts from one of the top manufacturers.

Other tight squeezes include the alternator and the headlight, and the throttle body and the fuel filter. Using the shortest Prelude alternator belt possible will create ample clearance by necessitating that the alternator be swung away from the headlight, back toward the engine. In any case, the original H-series belt won't work.

Except for having to be very careful where you drive because of the lack of clearance between the oil pan and the pavement, all other clearance issues are now taken care of. And, oh yes, you might not want to take that ground clearance issue too lightly, as many an oil pan has been smashed by speed bumps and potholes.

Notice how close the engine sits to the shock tower of the Integra. You can get plenty of clearance by cutting the valve cover off on the end.

Suspension and Axle Guidelines

Moving on to the underside of the vehicle, you'll find the axles are next on the agenda. A couple of solutions are available for your axle problems; we'll start with the easiest. If you use a '90–'97 Accord manual-transmission intermediate shaft, you'll find that a pair of '90–'93 Integra axles will slide right into place. A second solution involves taking the proper measurements and ordering custom-built axles to fit for the H-series half shaft.

Many aftermarket axle manufacturers produce units of much higher quality that aren't prone to breaking. The Drive-

The half shaft you choose will determine which axles you'll need. The female end of this intermediate shaft determines what axle to look for.

HCP Engineering

Since 1996, HCP Engineering has specialized in Honda and Acura engine modifications and engine swap mount kits. The EF-18 bolt-in kit developed by owner George Hsieh in 1996 was the first ever engine bolt-in kit to be introduced to the Honda marketplace. This idea inspired many other bolt-in kit designs, and created a whole new market.

HCP Engineering has been around for years now and takes pride in testing all of its products on the racetrack. The engineering department continues to do extensive research and development for each of its engine-mount kits. All HCP Engineering motor mounts are specifically designed to place the engine in the factory location of the vehicle, without compromising the drivability or integrity of the chassis. And by placing the drivetrain farther back than stock in some cases, HCP Engineering allows for the most optimal of axle angles, resulting in far less axle breakage.

The folks at HCP Engineering have also considered motor geometry and oil flow to allow for the utmost in engine reliability. Accessory pumps and air compressors haven't been forgotten about either, so there's ample clearance to run power steering and air conditioning when applicable. HCP Engineering has also considered the amount of hood clearance for each engine, while not compromising ground clearance in any way.

According to HCP Engineering, its engine mounts are the only mount kits on the market today using real automotive-grade urethane pieces supplied by Energy Suspension. The urethane pieces have specific durometer and density settings that allow for the utmost in strength, with minimal chassis vibration. The motor-mount kits and other products are professionally powder coated with black wrinkle paint. Billet aluminum motor mounts are now in production and will soon be available for most applications. New designs are always being tested and developed to maintain HCP Engineering's renowned cutting-edge status. All HCP engine mounts and urethane components are backed by an exclusive lifetime warranty.

HCP Engineering Honda engine-swap products include:
'88-'91 Civic-to-B-series mount kit
'88-'91 Civic-to-B-series air-conditioning compressor bracket
'88-'91 Civic-to-B-series shift linkage
'88-'91 Civic-to-B series throttle-cable adapter bracket
'88-'91 Civic-to-H-series weld-in mount kit
'92-'95 Civic and '94–'01 Integra-to-H-series mount kit
'92-'95 Civic and '94–'01 Integra hydraulic-to-cable transmission-adapter kit
'96-'00 Civic-to-H-series mount kit
'96-'00 Civic hydraulic-to-cable transmission-adapter kit
'96-'00 Civic to H series front crossmember
'96-'00 Civic driver-side replacement mount
'92-'00 Civic to H series air conditioning compressor bracket

Customized engine wiring harnesses and OBD adapter harnesses for most transplants.

shaft Shop is just one company that offers upgraded axles for the Integra to H-series conversion. You'll find that these are the same axles used on the '92–'95 Civic to H-series conversion as well. After you've finished the driveshafts, the suspension components can be reinstalled in the same manner that they were removed, without any modifications.

From Shift Rods to Shift Cables

Just when you thought you were done underneath, there is still plenty of work to be done with the shifter mechanism. Perhaps the most difficult and misunderstood part of this whole transplant, the shifter-assembly installation has been performed in at least three different ways. Fortunately, there is only one right way if you want your transmission to shift properly and the interior of the car to go back together like stock.

The first step in doing this correctly is selecting the right cables. Cables can be sourced from any '90–'97 Accord or '92–'01 Preludes. Although many of these cables are different from each other, they're all compatible with these H-series transplants. You must be careful to not mix and match parts between the two different styles of shifter cables. There are slight differences that make certain Prelude cables incompatible with certain Accord shifters. If possible, try to purchase a complete setup if one doesn't come with your engine.

To begin the process, remove the center console and pull the carpet out of the way. The original oval-shaped shifter hole needs to be cut into a square for the

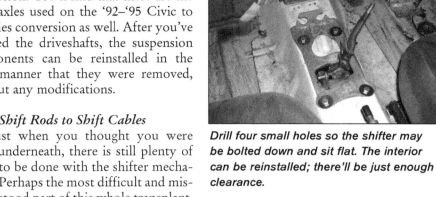

Drill four small holes so the shifter may be bolted down and sit flat. The interior can be reinstalled; there'll be just enough clearance.

Prelude shifter assembly to lay flat.

Use a reciprocating saw to make the necessary cuts from inside the vehicle. Be sure to remove only a little bit at a time, testing the fit of the shifter assembly in between cuts. With the shifter assembly lying flat, mark the four bolt holes on the chassis corresponding with the four holes on the shifter. Next, fasten the assembly to the car with some 8-mm bolts and nylock nuts. This will take two people, so make sure you have some help nearby.

Drill a hole exactly 18 inches forward from the center of the shifter from underneath the vehicle. The shifter cables can be routed through this hole. Watch for airbag wiring when drilling, as it runs right in that area.

When installed properly, the cables will run from the shifter assembly inside the vehicle until they slide through the hole in front of the shifter. In lieu of drilling holes and mounting the shifter assembly up top, a special mounting box

from Place Racing may be installed underneath the vehicle. Mounting the cables on the underside will eliminate the need to route anything inside the vehicle. Either way is acceptable.

With whichever mounting method you choose, the cables must also be routed over the rear crossmember and above the rear engine bracket before finally being attached to the transmission. When connecting the cables to the transmission, make sure that you have all of the proper washers and cotter pins, as these are usually missing from junkyard engines.

Back inside the vehicle, the center console needs some minor plastic trimming before being reinstalled. Before installing the console, double check that the shifter is able to go into all of its gears and that it's not hitting the chassis because the hole was cut too small.

Hooking up the clutch line is the last item to be addressed for the transmission. Although, again, there are several ways to do this, one way in particular proves to be the best. Rather than bending and tweaking the original hard line to fit, you'll need to get a little creative on this one. Simply disassemble the Integra or Civic clutch line at the connection on the frame right next to the right-side engine mount. Install a steel braided clutch line of the proper size and dimensions. Use an approximately 6-inch -3 steel braided hose to splice the new line in place. A couple of 10 mm to -3 pipe adapters with 90-degree hose ends will need to be added to the Prelude slave cylinder and the end of the original clutch line.

Spackman cuts out the appropriate square-shaped opening for the shifter plate with a reciprocating saw. Be careful not to cut away too much.

Notice the yellow SRS airbag wiring inside the small opening for the shifter. Hit one of these and you're in trouble. All of the SRS wiring is the same color and is very time-consuming to fix.

This customized hydraulic clutch line was made out of steel braided fittings and line from Earls Plumbing. Metric-to-AN adapters make it possible.

Don't forget to bleed the clutch after you've finished. One of the last items that needs to be done is the installation of the throttle cable. You need either a cable from any '98–'01 USDM Prelude or a special throttle cable bracket from Place Racing to work in conjunction with your stock cable. In this case, it's just a matter of personal preference.

Fuel-Injection Walkthrough

When dealing with fuel systems on any H-series transplant into a Civic or Integra, a bit of modifying must be done. The first problem that the installer will encounter with some mount kits is the interference between the fuel filter and the throttle body. In these situations, a special bracket must be installed to relocate the filter out of the way and fasten it properly to the frame. You can pick up one of these from Place Racing if you don't wish to fabricate one yourself. Although many folks unbolt the fuel filter from the firewall and just leave it dangling, this certainly isn't a safe idea.

When dealing with the H-series engines, often you'll find that the fuel injection feed port is on the opposite side of the engine bay. Rather than lengthening the fuel injection feed hose, install a fuel rail from a USDM '92–'96 H22A to fix the situation. The Integra

and certain Civic fuel injection feed lines are incompatible with the new rail, so the line must be swapped for the appropriate unit. One from the Prelude will work just fine.

Radiators, Hoses, and Fans

Moving on to other fluid systems, if you have an Integra, you'll find that the stock radiator not only fits perfectly, but also, thanks to its larger dimensions, will keep the new engine plenty cool. Of course, due to their smaller size, Civic radiators will fit just as well, although the thicker twin-core units are recommended. These can be acquired from the '94–'97 Del Sol DOHC VTEC and '99–'00 Civic Si.

You'll need to add a thinner cooling fan on all radiators to accommodate for the slave cylinder, which would otherwise rub against the fan during engine movement. A push/pull fan will allow you to mount the unit on the opposite side of the radiator. Be sure to read the wiring instructions on these fans before installation, as the reversal of the two fan wires will cause it to rotate backwards, resulting in overheating.

To finish off the cooling system, two radiator hoses and two additional water hoses need to be modified and reattached. The first is the hose exiting

You might want to install this heater hose before installing the H engine. It's tough to get back there once the rear bracket and intake manifold are sitting right in your way.

the water pipe on the rear of the Prelude engine, which needs to be routed to the water valve assembly located on the firewall. A piece of heater hose cut to length with adjustable hose clamps will suffice. Next, the original Civic or Integra inlet water hose can be attached to the side of the H-series cylinder head water port with the original clamps.

Upper and lower radiator hoses from a '92–'96 Prelude will fit perfectly once the top one is trimmed to size. Don't forget to use adjustable-tension hose clamps to help adapt the smaller water ports of the single-core Civic radiator to the larger Prelude hoses.

You Can Even Keep the A/C

If you wish to retain your air conditioning or power steering, be forewarned that you have your work cut out for you. Since the engine is already an extremely tight fit, you'll be hard pressed to get the components in place. Luckily, the special brackets from Place Racing, Hasport Performance, and HCP Engineering will all allow you to install the original Civic or Integra A/C compressor. What they won't allow you to do is reuse the original A/C hoses. Although the original hoses will bolt up into place, they'll eventually burn from being in constant contact with the radiator. This is a dangerous situation that you can avoid by taking measurements and having custom lines manufactured.

Once the lines are fabricated and installed, you'll find that the condenser fan isn't going to cooperate either. If you have an Integra, the lower housing of the

Here's what not to do with the fuel filter. Although it must be relocated, leaving it dangling like this is dangerous and could cause a real problem. At the very least, fasten it down with a large zip tie.

condenser fan will come into direct contact with the compressor, so the lower portion will need to be trimmed quite a bit to fit into place. An easier solution is to install a smaller fan like the one used on the radiator and wire it up on the opposite side of the radiator like the original condenser fan. This step isn't necessary on the Civic, but it will make for a better fit because of its compact size compared to the stock unit. The reason for the Integra's clearance situation is that the Integra's radiator is twice as wide as the Civic radiator.

Power Steering Tech

Moving on to power steering components, you'll agree that the situation is a little bit more straightforward. Let me put it this way: if you don't have a Hasport Performance or Place Racing kit, you aren't keeping your power steering. Thanks to these companies, the H engine will be offset to the passenger side of the vehicle just enough to provide room for the power steering pump on the driver side of the engine. This installation can be completed with the appropriate '97–'01 Prelude power-steering pump and bracket, as well as the factory belt. However, modifications must be made to the power-steering line.

The section of line that houses the fitting for the pump must be cut off and replaced with a section from the corresponding Prelude. This can be done by brazing or by an experienced TIG welder. Due to the extreme pressure, you'd better make sure that this is sealed

up good. If fabricating a homemade pressure line isn't your thing, the same folks who make your A/C lines can make a custom line. If this section of hose isn't swapped out, the line will be unable to bolt up to the Prelude pump. Once it's bolted in place without any leaks, the power steering will function as normal.

Other Concerns

As you'd probably expect, adding 185 pounds to the Civic chassis by means of an H series engine causes some adverse side effects. Even the additional 85 pounds added to the Integra chassis will be noticeable. The first item that should be addressed when swapping in such a powerful and much heavier engine is the braking system; this is especially true for Civics that aren't equipped with four-wheel disc brakes. This situation should be taken care of at the time of the transplant. Integras are a little bit better off as they're all equipped with four-wheel discs. At the very least, it's recommended that the pads and rotors be in good working order for the DC folks. As for the Civic owners, the '94–'01 Integra front and rear suspension and brake system are direct bolt ons. It's important to look into this option, as it would be in the best interest of your safety.

Other adverse effects due to the weight change will be noticed in the vehicle's handling characteristics. Negative effects of understeer will be unavoidable. To counteract this problem, Progress Suspension has a line of

specially designed lowering springs just for vehicles with engine swaps. One of their larger diameter racing rear sway bars will help alleviate the situation even more. Although the understeer cannot be fully eliminated, measures may be taken to make it very unnoticeable.

Final Words for the H Series

After all is said and done, if a competent swap mechanic does the transplant and a quality engine mount kit is used, it can be pretty darn reliable. However, it will be nowhere near as trouble free as a B-series swap, which is evident in a few of the H series' downsides. For starters, the axles still tend to be the weakest link of these two chassis. With most street-racer-type folks, replacing the axles on a bi-monthly basis isn't unheard of. The problem lies in the poor axle angle in conjunction with wheel hop caused by the frequent loss of traction. Lost traction is an everyday occurrence for some of the lighter CX and VX Civics, even if it isn't raining.

Other adverse effects include poor ground clearance and a lack of space under the hood, prohibiting many future upgrades. With all of these negative side effects, it's a wonder that anyone would even want to do an H-series swap. It all comes down to the bottom line: it's fast. I mean after all, that's all that really matters when you have the undeniable need for speed. Although this isn't the type of car that you'd feel comfortable driving your grandma around in, it certainly fulfills its purpose: pure acceleration.

This is right where you want to cut the power steering line if you plan on keeping it. Once it's cut, the piece on the left can be discarded for a similar piece from a Prelude.

This larger diameter anti-sway bar on the left from Progress Suspension is necessary for all Prelude engine transplants. It's a good investment because it helps to alleviate understeer.

Once you get it finished, you'll find that the Prelude engine sits quite snugly under the hood of this Civic. There will even be plenty of room left over for a custom header such as the one shown here.

1996 TO 2000 CIVIC

The introduction of the '96 Honda Civic unveiled a body style that was very different from the typical mainstream sport compact. With its ultra rounded exterior and larger than life headlights, it took some getting used to and a while for Civic lovers to come around to these new bodylines. Now considered by Hondaphiles to be the most aesthetically pleasing body style of all Civics, the sixth-generation chassis has sure come a long way.

Civic Offerings

Manufactured from 1996 to 2000, the EJ/EM chassis Civic (often referred to in the EK, JDM form) is going to be on the expensive side due to its relatively young age. Available in hatchback, two-door, and four-door form, this particular Civic has a body style to suit anyone's needs.

This available Civic offers trim levels from base to luxury include the CX, GX, DX, HX, LX, EX, and Si. Amenities include air conditioning, power steering, compact disc changers, moon roofs, and VTEC engines, which explains why the price ranges significantly between the trim levels. Hatchback CX models are the lightest, at 2,200 pounds. They're usually preferred among engine swappers due to their lacking anything other than the bare essentials. Top-of-the-line models such as the '99–'00 Si are the most costly, being that they're standard equipped with the 160-horsepower

The guys at Import Life remove the old D series out of this '96 Civic, making way for a more powerful B16A.

This '96 four-door Civic LX is a perfect example of an enthusiast trying to have the whole package. The owner of this Civic isn't satisfied with just exterior enhancements; he wants an engine swap to back it up.

These original D-series engines are pretty much worthless. They have very little power to begin with, and if you try adding a turbocharger or nitrous oxide you'll end up with a melted piston in a hurry.

B16A2 engine. Although not a recommended vehicle for an engine swap, the Civic Si shares all of the same chassis configurations as the more lowly Civics, making it swappable as well.

The '96–'00 Civics not equipped with rear disc brakes may find upgrades from the '99–'00 Civic Si, or any '94–'01 Integra for that matter. This upgrade is a straight bolt-on procedure, but remember to swap the emergency brake cables at the same time, as they're necessary.

Front brakes may be swapped in favor of larger third-generation Integra GSR or Type R units as well. You'll need the knuckles, wheel hubs, calipers, rotors, pads and brake lines for a proper conversion.

Suspension upgrades are no further away than the folks at Progress Suspension. They design springs especially for vehicles receiving engine swaps, so they can provide you with the components that will help eliminate understeer gremlins from the heavier engines.

In addition to the B16A2 found underneath the hood of the Civic Si, several other engines are available in this particular Civic chassis. One of the most powerful SOHC VTEC engines, the D16Y8, can be located in the engine bay of all '96–'00 EX models. Less powerful engines stamped D16Y7 can be found in the likes of the CX, DX, and LX models, while the D16Y5 can be found in all HX models. The Civic GX features a specially designed natural-gas engine and

should be avoided concerning engine swaps. Any of these Civics except for the GX and Si will make perfect transplant candidates, as their engines are obviously lacking in the power department.

OBD Tech

When dealing with electronics on the '96–'00 Civic, things are relatively simple thanks to the fact that they're all OBD II. Although two different versions of OBD II plugs exist, the systems use the same sensors and same emissions components. Honda changed plug styles in '99, so the '96–'98 harnesses are somewhat different those in '99–'00. Naturally, it follows that '96–'98 ECUs and '99–'00 ECUs are incompatible. On all '96–'00 EJ/EM Civics, the engine harness actually protrudes into the vehicle, making its way underneath the dash to ultimately plug into the ECU.

The ECU is hidden behind the passenger-side kick panel on the floor. This Civic engine harness is entirely different from all the others in this book. Although the harness still needs to be removed from the vehicle for the swap, any conversions that would normally need to be made underneath the dash can now be done mostly on the workbench.

When selecting an OBD II ECU for the engine swap, it's important to pick one that corresponds with the OBD II type of the vehicle being used. Failure to do so will result in unnecessary wiring to get the harness to accept the mismatched ECU. Since the vehicle is already OBD II, either USDM or JDM OBD II ECUs and engines may be used. Using the '99–'01 USDM Integra ECU will require disabling the built-in anti-theft system. Failure to do so will lead to a deactivated fuel pump. OBD I conversions can be made by adding an OBD adapter harness from either Place Racing or Hasport Performance. Plugging one of these in will allow you to use any OBD I computer. Be sure to specify which OBD II type your vehicle is before ordering anything.

B-Series Engine Swap

You can thank Honda's designers for allowing you to fit any B-series

The engine harness on this Civic makes its way into the vehicle through a hole underneath the battery. This must be pulled out before the engine can be removed.

engine relatively easily under the hood. Part of the reason it's so easy is the fact that several JDM and EDM '96–'00 Civics were equipped with a B16 of some sort right from the factory. We didn't receive our version of the B16A in the sixth-generation Civic until the release of the '99 Civic Si. By that time, B16A swaps into sixth-generation Civics were already being done by the multitudes. Anytime you have a JDM or EDM version of a USDM vehicle with a bigger, more-powerful engine, it stands to reason that our USDM version should be able to receive that same engine, or at least one that is similar to it. Since B-powered Civics exist in other parts of the world, half of the installation process lies in finding the proper parts to make things happen. With the right components, this Civic-to-B-series transplant is just as easy as performing one on the older EG Civic. It's a straight bolt-in affair to say the least, and even an amateur Honda swapper should be able to complete it in a weekend. Basic mechanical skills will be necessary, of course, but overall, this is one of the easiest transplants you're likely to encounter.

B-Series Engines and Transmissions

Since all '96–'00 Civics are equipped with a hydraulic-style transmission and OBD II electronics, certain engine swaps will make life easier concerning the install. Two B-series engines should come to mind if you're looking for a

USDM OBD II engine: The '96–'97 Del Sol B16A3 engine, and the '99–'00 Civic Si B16A2 engine.

OBD II B16A engines make a perfect donor engine for this chassis, as it's totally compatible in terms of mounts and electronics. This one is from a Civic Si.

Although the earlier-model Del Sol engines are identical and will work, they're equipped with OBD I electronics and will either require a wiring conversion be performed on the vehicle or that minor electrical changes be made to the engine to accommodate an OBD II ECU. Although backdating the ECU to accommodate an older engine is very common, for simplicity's sake it's recommended that an OBD II engine be considered first. Compatible JDM B16As include those in the '96–'00 Civic SiR and SiRII, as well as the '92–'97 CRX del sol SiR. Since only the '96–'97 CRX del sol versions are OBD II, they'll be your only JDM choices if OBD is a deciding factor. An EDM version can be sourced from the '96–'00 Civic VTi. A final OBD II B16 engine can be found in the '98–'00 JDM Civic Type R. This would make an excellent addition to any OBD II Civic.

Stepping up to the larger, more-powerful members of the B-series family, several B18 and B20 engines will work the '96–'00 Civic. If you're going the non-VTEC route, the '94–'01 USDM Integra RS and LS B18B1 engines will work perfectly. JDM versions can be found in the Integra ESi base model as well. Other non-VTEC engines include both the B20B and B20Z, available in '97–'00 USDM CRVs and the '99–'00 JDM S-MX, respectively. Of all of these non-VTEC

engines, only the '96 and newer versions are equipped with OBD II.

If horsepower is one of your biggest concerns for an engine swap, then you'll at least have to consider the B18 VTEC engines. There are several available, and we'll start with the USDM models first. The '94–'01 Integra offers a few different variations of the B series, two of them being VTEC. The B18C1 is available in all eight years of this Integra's production in the form of the GSR. A

more powerful version of the B18C is available in the '97–'01 Integra Type R, excluding the 1999 model year. This B18C5 is the most powerful and most expensive B-series offering that is available in the domestic market. Moving on to JDM engines, the B18C is also available in the '95–'97 Integra SiR-G. The Japanese version of the Type R (also known as a B18C) can be found in the '94–'01 Integra Type R. Other available B18 VTEC engines may be acquired

Not That Kind of B Series

You hear a lot about the interchangeability in B series components. With all the parts swapping going on between the B16s, B18s, and B20s, it has become a general assumption that if it's a B series, then it must fit. Even though this is true to some extent, there are a couple B-series engines floating around that share no more with the engines we generally think of as B series than the B in their name. I'm talking about the B engines of the '88–'91 Prelude.

Commonly referred to as the B20 and B21, these engines couldn't be further from the Integra-style B-series engines if they tried. They have different bolt patterns on almost all of their components, so the common B-series interchangeability in

cylinder heads, transmissions, and engine internals is next to impossible. Although B16A VTEC cylinder heads can be retrofitted onto these blocks, it certainly isn't a bolt-on process. With all the machining and drilling needed for the adaptation to take place, you sometimes have to ask the question, why? Seriously, with B18A blocks at a dime a dozen, there's no need for a conversion like this. Moreover, since these were built before the mass-production of VTEC, and before the arrival of the H-series in the Prelude, these engines never really received any attention from the aftermarket. Regardless of what you do, avoid these engines if at all possible, because they will be no help for any of the engine swaps in this book.

These B-series engines found in the '88–'91 Preludes are not only completely different from the Integra-style B series, they're also pretty much hopeless as far as performance is concerned.

B20B engines are just as compatible as any other OBD II B series engine would be. Strapping it up to any B-series transmission will make for an easy install.

This late-model hydraulic transmission can be identified from this angle by its mount, the starter, and the hydraulic line bracket on top. Any year will suffice, as long as it's a hydro unit.

These JDM computers look a little bit different from the USDM versions. Their square shape makes them easy to identify in contrast to the rectangular units we have.

from the EDM Integra GSR and Type R, although these are much harder to find here in the United States. The '96–'01 Integra engine is the only one available with OBD II electronics.

When deciding upon the proper transmission to use, the rules are simple. If you want to make things easy, then go with the later-model hydraulic unit. You can find these partnered with any of the B-series engines listed above. If you don't mind a little extra work and are looking to save some money while you're at it, then the older, less-expensive Integra cable transmissions will work too. Of course, using one of these will require you to add a hydraulic-to-cable adapter kit from HCP Engineering, Hasport Performance, or Place Racing. The nice thing about the hydraulic-style transmissions, on the other hand, is how well they fit into place without any aftermarket mounts. If a stock appearance is your goal, or you simply don't want a transmission built in the late 1980s, then the hydraulic unit is your best bet. For best results, it's suggested to seek out the rare limited-slip differential versions available from vehicles such as the Type Rs. Geared strictly for pure acceleration, these transmissions may not provide you with the best gas mileage.

ECUs and Wiring for the B Series

Of the many different OBD situations possible for this transplant, only two should be considered. The first and easiest choice is to keep the OBD II sys-

tem. This will require a 1996 or newer OBD II engine and ECU. You can plug the new ECU right into the original location to keep wiring at a minimum and ensure a speedy swap process. Just be sure to use an OBD II computer that corresponds to the OBD II plug type of your vehicle.

The second choice would be to purchase any of the engines listed above, but use an OBD I ECU. You'll need a Place Racing or Hasport Performance adapter harness for underneath the dash to be able to plug the new ECU into place. You'll also need to do some basic modifications to the engine harness; these are explained below. Unless an ECU is only available in the OBD II form, or state laws do not permit you to use one, then for performance reasons an OBD I conversion is going to be the way to go. Ultimately, your choice should made by how much work you want to do, how much you want to spend, and what parts are readily available to you.

There is a wide variety of computers available depending on which OBD ver-

sion you choose. Non-VTEC OBD I folks should using either the '92–'93 Integra RS, LS, or GS PR4 ECU, or the P75 ECUs found in the '94–'95 Integra RS and LS.

The OBD II swap will require the totally different P75 ECU from the '96–'98 Integra RS and LS. VTEC ECUs are often interchanged according to availability and ability to be reprogrammed, so matching corresponding ECUs to engines isn't always as important.

OBD I B16A engines generally use the '94–'95 USDM Del Sol VTEC ECU, otherwise known as the P30, but JDM versions of this ECU will work equally well. Other OBD I B16A-compatible computers include the P61 unit from the USDM '92–'93 Integra GSR and the JDM PR3 (not the OBD 0 unit) from the '92–'93 Integra RSi or XSi models.

OBD II B16A ECUs include the P2T unit from the '99–'00 USDM Civic Si, as well as any of the P30 computers from the '96–'97 Del Sol VTEC. Keep in mind that even though they're both OBD II, the Si ECU will only work on the '99–'00 Civics unless you perform a

Hydro-to-Cable Conversions

For those who choose not to use the newer hydraulic B-series transmission, you're in luck. Rather than fabricating a custom transmission bracket, several manufacturers offer adapter kits that retrofit these older and less-expensive gearboxes into the late-model chassis. By purchasing a conversion kit from Hasport Performance or HCP Engineering, any of the '92-'00 Civic and '94-'01 Integra B-series transmissions can easily be bolted into place. Simply install the bracket and mount it onto the transmission with the necessary hardware. This way, you can maintain all of the vehicle's original hydraulic operating system, and allow it to function with the cable transmission.

wiring conversion. B16 Type R engines should use the PCT ECU found in the JDM Civic Type R, of course.

When installing a GSR engine, the P72 is your only option if you want to keep the intake air bypass system in the B18C intake manifold. P72s exist in both USDM and JDM OBD I and OBD II vehicles. OBD II Integra Type R engines will require the P73 ECU found in any 1996 and newer Type R, JDM, or USDM. If all else fails, the '92–'95 USDM Civic EX and Si ECUs can be used as well. With the P28's low price and the ability to be reprogrammed to DOHC VTEC status, it remains among one of the most popular choices for any DOHC VTEC swap.

When selecting USDM OBD II Integra ECUs, avoid the '99–'01 units, as they're equipped with a vehicle immobilizer device. Although this anti-theft deterrent can be removed, it will cost you extra cash to have it done, and there aren't many folks who know how to do it.

Regardless of which computer you choose, in most cases you'll need to perform some wiring modifications to the Civic engine harness. With the original engine harness removed from the vehicle, now is a good time to place it on the new engine for a test fit. This will give you a good understanding of what must be lengthened, shortened, or left alone. Providing that the OBD status of the vehicle is to remain OBD II, minimal wiring modifications will need to be made. If you're using a non-EX or Si engine harness, you'll probably have to swap the idle air control valve plug. Due to the differences between connectors, the plug must be swapped out for the connector found on the donor engine's wiring harness. This must be performed on both VTEC and non-VTEC engines, although the VTEC engines will require even more wiring.

EX vehicles are the lucky ones, since they only require you to lengthen a few wires to ensure a proper fit, and no wiring modifications are needed for the engine harness, including VTEC. Just plug it into any ECU that corresponds to the OBD II version of your vehicle and you're done. What's nice about this EX engine harness is that it can be transferred

over to other vehicles. In fact, any of the '96–'98 EX or HX (EX alternative) harnesses can be swapped over to any '96–'98 Civic. The same thing goes for the '99–'00 and '99–'00 EX and HX harnesses. In fact, if you have a '99–'00 Civic, you can also use the '99–'00 Civic Si harness, which will also be a plug-and-play affair. When using the Si harness however, you'll notice an extra wire that remains in the harness side of the distributor plug. Simply unpin this wire or cut it off; it is unnecessary.

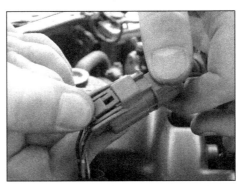

As with any transplant, some connector swapping will be inevitable. You can either re-pin the plugs or cut and solder the wiring.

For those of you who aren't interested in spending any more cash on a new EX or Si harness, of course, the existing engine harness may always be modified just as in previously mentioned swaps. In addition to the IAC valve plug modification, the crankshaft position sensor will need to be modified as well. The electrical part of the crankshaft position sensor must be swapped out for the plug on the donor engine's harness.

Other wiring modifications will be necessary for any VTEC engines being used. In order for VTEC to function properly, you'll need to add the VTEC wiring to the engine harness. This includes the knock sensor, VTEC pressure switch, and VTEC solenoid. On select JDM engines only, you don't need to wire the VTEC pressure switch as long as you're using the corresponding JDM ECU. If you're using a non-JDM ECU, the entire VTEC solenoid unit must be replaced with a unit that has provisions for a pressure switch. To keep things simple, try to use the computer

that is sold with the engine if possible.

To wrap up the OBD II wiring information, we need to look at what's necessary for using a '99–'01 Integra ECU on a '99–'00 Civic. If you go with this ECU, and you've managed to have the anti-theft immobilizer removed, then you'll also need to make a minor modification to the ECU-side of the engine harness. In order for the Civic fuel pump to function with this particular ECU, the wire from pin A16 must be removed and placed in the A15 location. If you do not have the means to have the immobilizer removed, an adapter harness may be purchased from Hasport Performance to convert your vehicle from the other plug style of OBD II. This will allow you to use the earlier-style '96–'98 Integra computers.

For those who hope for more horsepower and refuse to remain with OBD II, an OBD conversion will be necessary. In addition to adapting the harness for the ECU, standard VTEC modifications must be made, as well as lengthening and shortening the harness as necessary. Final wiring modifications include swapping the dual distributor connectors off the Civic engine harness for the single-plug connector off the donor engine's harness. The injector plugs must be swapped out for older units as well.

While on the topic of OBD converting, if you want to use an OBD II engine and backdate the electrical system to OBD I, things aren't going to be too different. In fact, they'll be just the same, except for the fact that you won't have to modify the distributor plug or the injector plugs on the engine harness this time. Simply install the adapter harness of choice, plug in the ECU, wire VTEC, and retrofit the harness to the new engine. It's as simple as that.

Installing the B Series

Once the wiring portion of the swap is figured out, the harness can be installed onto the engine. Next, you need to select the proper engine mounts and bolt them into the car. The rear engine mount can remain intact, but using a rear engine bracket from a '99–'00 Civic Si is mandatory.

With the B-series swap, this rear mount and heater hose should be left on the vehicle. Since they're going to be reused, don't waste your time taking them off.

Late-model Integra axles are generally used on these swaps even though several other early Integra and custom combinations are possible.

There are no compatible Integra or Del Sol mounts for this swap; they're all too long. The left-hand engine bracket must be replaced with a unit from the '99–'00 Civic Si or '94–'97 Del Sol DOHC VTEC. Again, the Integra pieces aren't compatible with this chassis. With this Si or Del Sol bracket, the rubber mount attached to the frame rail and the corresponding bracket may both be used again. The top transmission mount as well as the bracket can also be reused, providing that they come from a five-speed Civic. To finish off the mounts, the front transmission bracket can be sourced from any '94–'01 Integra, '99–'00 Civic Si, or '94–'97 Del Sol DOHC VTEC.

Lastly, if you want to retain the air conditioning, you can use the front left bracket from any of the vehicles just mentioned. These last two brackets will slide into corresponding rubber stopper mounts already underneath the frame. It's very important that these original stopper mounts are reused, as the Integra units are significantly different. Hook up the left side mount first, and then slide the rear engine bracket into place, before bolting on the transmission side top mount. With all those in place, you can tighten up the front mounts. If the proper mounts are used to begin

Although the Civic transmission brackets are the same as those for the B series, the automatic version of the Civic bracket on the right must be replaced with the manual version.

with, most folks will never even know the engine has been swapped out. In addition to the older EG B-series swaps, these are the most factory appearing transplants of any Civic or Integra. Thanks to its OEM fit, there won't be any type of clearance issue.

Suspension and Axles Simplified

As you'd probably expect, reinstalling the suspension and the axles isn't too difficult on this swap. In all cases, the original factory suspension will bolt right back into place, as will any strut bars added underneath the hood. Axles from the '94–'01 Integra can be used, providing that you also use the corresponding intermediate shaft.

Other options include using the '90–'93 Integra axles and intermediate shaft. This will work as long as you replace the axle seal on the intermediate-shaft side of the transmission with the earlier Integra-style piece. Del Sol DOHC VTEC and '99–'00 Civic Si parts combinations will work as well.

Shift Linkage and Cables

To finish things off underneath the car, you'll need to add a shift-linkage assembly from a '94–'01 Integra or '99–'00 Civic Si. Avoid the Del Sol unit, as it isn't the right size for the Civic chassis.

If you use a hydraulic-style transmission, you'll find that the clutch line will reattach to the new Integra slave cylinder without a hitch. If you happen to purchase a late-model B16A from a Civic, you might be lucky enough for it

If you use an OEM throttle cable, you'll find that it attaches to the pedal assembly, firewall, and throttle body all in an OEM fashion.

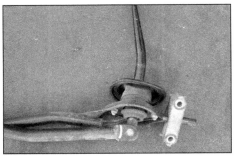

This shift linkage mechanism from the Integra will provide a factory bolt-on fit without any modifications. The Civic unit is worthless here.

to include the bracket that holds the clutch line to the starter. If not, an insulated hose clamp can be used in order to avoid unnecessary vibration of the hydraulic clutch line. In most cases, a throttle cable from a '97–'01 Integra Type R will work perfectly unless you're installing a GSR engine. In this situation, you need to install a cable from the '94–'01 Integra GSR. Place Racing also manufactures throttle cable brackets for some of these swaps, so it's easy to reuse the Civic cable.

Fuel-Injection Pointers

Moving on to the fuel systems, things prove to be reasonably straightforward here too. With a couple different fuel-hose banjo ends available in the Honda world, you need to make sure you have the same one before attaching the fuel line. When swapping in a late-model Civic Si engine, for example, you can reuse the original fuel injection feed line. For many other B-series engine transplants into this chassis, an injection feed line and the corresponding nut and sealing washers from any Integra must be installed. Failure to do so will result in inadequate fuel supply and poor engine performance. The complete set may be acquired from any late-model Integra, among other vehicles.

In most cases, the fuel-injection return line can be reconnected very easily. In some cases, the regulator outlet is somewhat larger and you'll need a step-down adapter to match up the two different sized hoses.

When performing this modification, make sure you always use fuel-injection

The fuel regulator outlet in this case is slightly larger than the return line found on the Civic. A reducer will need to be used in between the two hoses in order to make the connection.

line and clamps. The burst rate of normal fuel line is very low and has been known to crack and start fires on injected vehicles.

Cooling System How To

Thanks to the B-series engine's stock-type fit, the original Civic radiator and its cooling fan may be retained. The original bracket can be bolted back up to the hood support just as it was removed. Before filling up the radiator with coolant, you need to select and install the cooling hoses. As with any B-series transplant, the lower radiator hose found on the '94–'01 Integra GSR is usually your best bet. An upper radiator hose from the '97–'01 Integra Type R or '94–'97 Del Sol DOHC VTEC should be used on any B16A or Type R engine transplants. For B18C GSR and non-VTEC swaps, the '94–'01 Integra GSR upper hose is a perfect fit. If you prefer not to squish down the larger B-series radiator hoses onto the smaller water

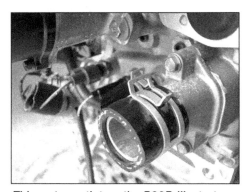

This water outlet on the B20B illustrates the larger diameter of the B-series radiator hoses compared to the Civic hoses. The Civic radiator has smaller inlets and outlets than this water neck.

ports of the Civic radiator, then you have another option.

Installing a twin-core radiator from a '94–'97 Del Sol DOHC VTEC or '99–'00 Civic Si will not only provide you with the larger inlets and outlets, but also far superior cooling capability. It might not be a bad idea to install one of these units anyway due to its ability to keep the larger engine that much cooler. To finish off the cooling system, attach the Civic water-inlet hose to the water-inlet pipe with the original clamp.

A/C and Power Steering Made Easy

If you want to retain air conditioning or power steering on any of these vehicles with the B-series swap, you'll soon see it won't be too hard. As far as the A/C goes, you just need to buy one special bracket. The Civic A/C compressor can be reused with a front engine bracket from '94–'97 Del Sol DOHC VTEC. Other brackets will bolt up to the block and attach themselves to the vehicles frame just fine, but the Del Sol bracket is the only one that also provides a home for the Civic compressor.

Unfortunately, the power steering isn't going to be that simple, but it's still relatively easy. It's just a matter of having the right parts. The '96–'98 pumps differ slightly from the '99–'00 pumps, so you might think that you have two options. Well you don't, because neither of them will work. You actually need to install the pump found on the '98–'01 Integra. This pump will allow you to reuse the Civic power-steering feed line,

This power steering pump is a straight bolt-on process if you use the proper Integra components. Use an Integra belt here as well.

---

and it will bolt up to any of these B-series engines with the proper brackets.

Lower power-steering brackets will remain the same for all B series, but the upper bracket will differ depending on whether or not the engine is anything but a B18C GSR or a non-VTEC B series. The B18C GSR engines require the upper bracket from the '94–'01 Integra GSR, while the LS and CRV engines need the bracket from the '94–'01 Integra RS and LS. Just because the pump is a '98 or newer unit doesn't mean the brackets must be as well. As for all of the other B16A and Type R engines, these require the bracket from the '97–'01 Integra Type R.

Possible Upgrades

With the exception of the '99–'00 Civic Si, none of the other EJ/EM Civics were fortunate enough to be equipped with rear disc brakes from the factory. Although they aren't an absolute necessity, rear disc brakes just might provide you with that extra bit of safety that you'll need when dealing with all this extra power. Fortunately, rear disc brakes can be swapped over from any '94–'01 Integra, as well as from the '99–'00 Civic Si. Just make sure that you install the emergency brake cables along with all of the other components. If front brakes are your concern, then

larger disc brakes may also be acquired from either of these same two vehicles as well. This might be an especially important issue if you happen to own a Civic CX with the small disc brakes up front and wimpy drum setup out back.

In addition to occasionally needing to upgrade the brakes to go along with an engine swap, it is also necessary to upgrade the suspension. Fortunately, in this case, the new B-series engine only weighs 100 pounds more than the original D series. With this slight increase in weight, negative understeer effects will be negligible. If adverse steering gremlins become apparent down the road, contact Progress Suspension for some of their purpose-engineered engine-swap springs and rear anti-sway bars to help alleviate the situation.

The B-Series Advantage

Thanks to the B-series factory appearance and fit, this is one of few transplants that will be maintenance free in terms of swap-related problems. With the engine position in Honda's desired location, axle problems will be rare, as will broken engine mounts. If you're planning to swap an engine into a '96–'00 Civic, the B-series should definitely be given strong consideration. Although it's a bit more expensive than the H series, and not quite as powerful, the B series

offers a few things that big brutes like the H22A cannot. These include a quiet and vibration-free ride, easy A/C and power-steering retention, and the ability to pass even the strictest emissions tests.

H-Series Engine Swap

If you happen to own a '96–'00 Honda Civic and the B-series engines just aren't powerful enough for you, then you really don't have too many options as far as other engines go. As with many unsatisfied B-series owners, the H-series engines are the next frontier in terms of larger-displacement engines and more horsepower. The sixth-generation Civic is one of the more popular vehicles to receive an H-series transplant, as it has just enough room underneath the hood for one of these babies. Although it isn't the easiest of swaps, with the proper tools and know-how, the EJ/EM-to-H-series swap can be completed in a couple of days. Unless you really know your way around a Honda, this isn't a recommended swap for first-time engine swappers.

H-Series Engines and Transmissions

For the sake of the electrical system, the ideal H-series swap into this chassis involves an OBD II engine. Unfortunately, this isn't practical for most OBD II Prelude and Accord ECUs. The number one reason that this isn't a good idea is the immobilizer system in the '97–'01 USDM Prelude ECU. This anti-theft device won't allow you to use one of these computers in any car other than the one that it originally came from. This is the result of its direct relationship to the ignition and the fuel pump. Although immobilizer systems can be removed from these ECUs, it isn't something you can readily have done, not legally anyway. In fact, even when the immobilizer has been removed, the Prelude and Accord ECUs never respond as well as they did in their original state. The bottom line is to avoid these computers if possible. If an OBD II must be used, then any of the '96 USDM Prelude or Accord computers and all OBD II JDM units will work without any modifications.

Similar to what you would find underneath the hood of the '99–'00 Civic Si, the B16A provides you with the most factory fit that you could hope for from an engine transplant.

This H22A engine has just been picked up from the wrecking yard. Notice how it hasn't been prepped yet. There are several hoses and lines hanging half cut off the engine.

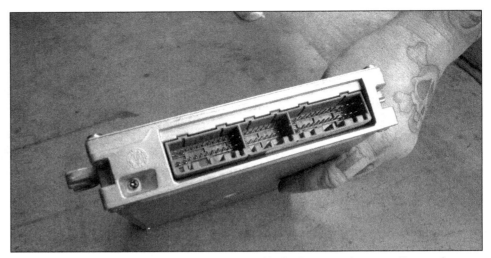

Ideally, this OBD I computer will provide you with the best results regarding performance. The OBD I units make the most sense since they can be reprogrammed easily.

When shopping for an H-series engine, the H22A Prelude VTEC engine is the most popular. The H22A is the most powerful four-cylinder Honda engine of its time, and surprisingly, it will fit very nicely into a sixth-generation Civic. The H22A that you're probably most familiar with can be found in the USDM '93–'96 Honda Prelude Si-VTEC.

This particular version of the H22A is stamped H22A1 on the engine block. Similar H22As can also be found in the '92–'96 JDM Prelude Si-VTEC, as well as the '94–'97 JDM Accord SiR. Except for other hard-to-find EDM models, this completes my list of a few recommended H22As. Although H22A-compatible ECUs are available for the 1997-model engine, remember that with the USDM units only the 1996 version isn't equipped with the immobilizer.

For those not interested in a VTEC engine, or who simply wish for a little extra displacement, look no further than the H23A. H23A engines are found in the '92–'96 USDM Prelude Si, as well as the '92–'96 EDM Accord 2.3i, stamped as the H23A2.

All of these engines use the '92–'96 hydraulic transmission that is compatible with all of the popular engine-mount kits. Although several different transmissions are available with various gear-ratio combinations, the outward physical dimensions are all virtually the same. Most folks will choose the gearboxes that are normally mated to the H22A engines because of their closer, racing-oriented gear ratios. Limited-slip

differential versions are available from many JDM junkyards if you look hard enough. In addition to the transmissions found in the VTEC models, a couple of other units will work too. You might not be aware that all of the '90–'97 Honda Accord transmissions are compatible, and will bolt into place just as easily as the Prelude units. Although much cheaper in price, these feature gear ratios that are far less desirable than even the H23A transmissions.

Of the remaining Prelude-type engines, most are 1997-and-newer versions that are equipped with the ATTS transmission. Although the ATTS was a major technological breakthrough for Honda, wiring up this additional computer that is dedicated to the transmission alone takes hours of unnecessary work. The fact that the older H- and F-series transmissions bolt right into place and work extremely well makes the ATTS units unnecessary to say the least. These older, non-ATTS transmissions are purely mechanical in nature, and thus do not require any wiring. Another downside to these 1997-and-newer engines is the fact that they're only equipped with the type of ECU that houses the anti-theft immobilizer system. If you purchase one of these newer engines, don't worry about either the computer or the transmission, as these will prove worthless soon enough. If an OBD II H engine is still desired, the H22A4 may be acquired out of the '97–'01 USDM Prel-

ude SH. The JDM '97–'01 Prelude SiR is a similar, more-powerful VTEC engine. Last, the only compelling reason for considering an OBD II H engine lies under the hood of the '97–'01 JDM Prelude S Spec. The H22A found there makes a convincing argument for the OBD II H series by thumping out 220 horsepower.

ECUs and Wiring For the H Series

Since most people will avoid OBD II ECUs, I'll focus most of my discussion of the electrical system to OBD I. An adapter harness of some sort must be used when using an OBD I Prelude or Accord ECU in the '96–'00 Civic. As in most cases, compatible plugs may be purchased and soldered into place under the dashboard, or a premade harness may be purchased from Hasport Performance or Place Racing. If you're skilled when it comes to automobile electronics, then the first way might be perfectly suitable for you. Before purchasing the plug-in adapter harness, it will be important to recognize the OBD type of ECU to be swapped. We already know that it is OBD II, but remember that the ECU plugs are different on the '96–'98 Civics and the '99–'00 Civics. For OBDII JDM ECUs, you'll need the '96–'98-style plugs.

With an adapter of some sort in position, you can start thinking about ECU selection and installation. P13 ECUs from the '92–'95 Prelude VTEC can be used on virtually all of the other Prelude

As with most Prelude swaps, you need to add a resistor box since these Civics aren't equipped with one. This 10-year-old Civic resistor box will work just fine.

engines listed earlier, providing you want an OBD I unit. Other compatible computers include the P72 ECU from the '94–'95 Integra GSR, the P30 ECU from the '94–'95 Del Sol, and the popular reprogrammed P28 from the '92–'95 Civic EX and Si. More often than not, you'll find that many folks with EK Prelude transplants are using anything but the P13 Prelude ECU. This can be attributed primarily to the fact that not many programs can be burnt onto the P13 computer board. When performing any type of non-VTEC H23A engine swap, the OBD I ECU of choice is the P14. These can be found in the '92–'95 USDM Preludes that are equipped with H23As.

Several wiring modifications will need to be performed for the computer to function properly with an OBD I Prelude engine and sixth-generation

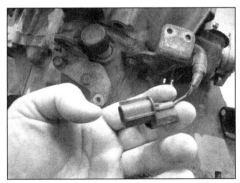

This reverse light connector found on the transmission must be swapped out for the Civic version. The Prelude connector is different and won't plug into the Civic engine harness.

Civic chassis. First off, the injector plugs will need to be swapped out for plugs found on the junk harness of the donor engine. Once soldered into place, the Civic engine harness must be retrofitted and tailored to size for the new engine. You can do this by plugging the harness into the fuel injectors and using that as a starting point.

From there, the intake air temperature sensor, starter solenoid, and alternator wires must all be lengthened for the connectors to reach the Prelude components. Other modifications include the addition of a fuel injector resistor box. Resistor boxes such as these may be found on older OBD 0 Civics and Integras, as well as most Preludes. A fuel-injector resistor box is only necessary when using a Prelude computer.

Once wired into place, the distributor plugs, the oil pressure sending unit

MAP Sensor Issues

This is a late-model MAP sensor assembly attached to the throttle body. Some other models have their MAP sensors on the firewall.

The wires are extended to reach the throttle-body-mounted MAP sensor.

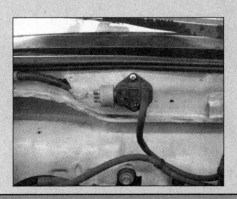

Honda implemented manifold absolute pressure (MAP) sensors into their vehicles in two different ways. Originally, MAP sensors were mounted onto the firewall with a special bracket. A vacuum line exited the sensor and attached to the throttle body to read its vacuum reference. With the introduction of OBD I and newer vehicles, the MAP sensors later became fastened directly to the top of the throttle body on most engines. The MAP sensor could now read the pressure directly from the hole it sits on. Both sensors work in much the same way and are surprisingly interchangeable between most vehicles.

With the introduction of the engine transplant came a set of issues involving the MAP sensor. The first set of problems arose when installing an OBD I or later engine into an OBD 0 vehicle. The older MAP sensor of the vehicle was completely different from the one from the donor engine, so one had to be chosen. The one fixed to the top of the throttle body is usually chosen. In this situation, you'll usually need to lengthen the corresponding wires of the vehicle harness to reach the new sensor.

A second scenario occurs during an installation where the donor engine doesn't have a MAP sensor. Several Prelude engines use the older firewall-mounted MAP sensor despite their newer OBD status. If you want to install one of these engines into a late-model Civic or Integra without the firewall-mounted MAP sensor, one will have to be acquired. Since many of the older Civics, Integras, and most all Preludes are equipped with this type of sensor, it shouldn't be too hard to locate one. A newer-style sensor could also be used, but they are usually difficult to mount because they were designed to sit in a special port on top of the throttle body.

The older style MAP sensors are the easiest to mount externally. Make sure the vacuum line connecting it to the intake manifold is in good shape; the computer needs an accurate reading to make the engine run right.

plug, and the reverse-light sensor plug must all be swapped out for Prelude harness plugs.

If you're using a P13 ECU, you must also wire the EGR valve and air intake bypass solenoid up to the ECU. You'll also need to modify all harnesses with VTEC wiring for those engines that are so equipped. The VTEC pressure switch and VTEC solenoid must be wired up to the ECU on all non-EX, Si, and HX models. A knock sensor must be wired into place for all H-series engine swaps as well. Note that the EX and HX harnesses are already equipped with one. External-coil Prelude distributors may be used, but they may also be converted to internal status by simply installing the proper coil and making the necessary wiring adjustments.

Installing the H Series

Assuming you have the whole wiring situation figured out, you can turn your attention to selecting and installing motor mounts. Just before installing the engine, you'll need to perform a couple of key modifications to the Civic. First off, the rear engine mount must be removed and replaced with a custom unit from HCP Engineering, Place Racing, or Hasport Performance. Each company's rear mount differs slightly and will install in a different way. Some require you to tweak the water valve on the firewall to free up some needed space, while others will simply bolt right into position.

This specially designed HCP Engineering driver-side mount is necessary for all H-series swaps. It fixes the all-too-common problem of broken mounts.

Speaking of the water valve, the water inlet hose must be removed from the valve at this time. Be careful not to damage or kink the delicate outlet pipe; use a knife blade to cut the hose off instead of pliers. A section of heater hose with an adjustable hose clamp should be placed back on the outlet at this time. This will be extremely difficult to finagle into place once the engine is set in. You can connect the other end to the engine later on.

If you're using any of these engine-mount kits, you'll need to bolt a separate bracket to the right-side frame rail before installing the engine. You can attach a corresponding mount to the top of the hydraulic Prelude or Accord gearbox of your choice.

Many engine mount kits are equipped with a custom front crossmember such as this unit from HCP Engineering. These are a good idea since they're designed to help stabilize the more powerful engine.

Attach a left-side engine bracket from any '92–'01 Prelude, and then attach the aftermarket driver-side mount of choice. At your discretion, the left-side rubber engine mount that attaches to the aforementioned bracket should be swapped out during this procedure. Although the original Civic piece will fit just fine, it's prone to breakage and should be replaced with a polyurethane version available from Place Racing and HCP Engineering.

Once you've done that, you can lower the engine into position at a slight angle with the transmission at the lower end. You can best accomplish this with an adjustable load-positioning bar placed on the end of the engine hoist. Attach the left mount, and then slide the rear bracket into position. The rear bracket from a '92–'96 Prelude with the manual transmission is your only option. Bolt this bracket to the rear mount and leave it unattached to the engine until after you've hooked up the two side mounts. Once you've bolted the rear transmission mount to the bracket, you can tighten the rear engine bracket from underneath. If your kit includes a front crossmember and front engine mount, now is the time to install them. These should usually be fastened to the existing bolt holes on the frame rails that house the vehicle's tow hooks.

If you use the HCP Engineering engine mounts, you'll have to remove or cut down the studs on the transmission. The mount will fail to seat down all the way, if the studs are left in place.

Once you've installed your H-series engine properly with any one of these motor-mount kits, you'll find that the '96–'00 Civic provides ample clearance in all directions. Although ground clearance is at a minimum, there is still quite a bit of room under the hood and around the engine bay. It would be wise not to lower the car any more than a couple of inches, because the extra weight of the engine will lower it even more once installed. This will bring the oil pan and transmission case all too close to the pavement, so watch out. In order to eliminate any future clearance problems under the hood, you should use the shortest alternator belt possible. This will help swing the alternator away from the headlight housing and A/C lines.

Key Points Regarding Suspension and Axles

After you've installed the drivetrain, the original suspension may be bolted back up into its original location. Since the Civics and Integras have a smaller wheel hub than the Preludes, you'll need some custom axles for this swap. Unlike Civic swaps that interchange Integra components on the front

Axles and Intermediate Shafts

Another confusing aspect of the engine swap process, the axles can pose quite a problem if you aren't sure what you need. The dilemma is that to the untrained eye, many Honda axles look alike. Heck, even to the trained eye, a lot of them still do. The color code markings that Honda uses to differentiate between their driveshafts are usually worn away, making them indistinguishable in a hurry. With so many different lengths, boots, diameters, intermediate shafts, and spline types, it's no wonder that finding the right axle combination can sometimes be so difficult.

Most engine swaps in this book will using factory Honda or Acura axles; some require minor modifications, and others just slide right into place. A few of the other remaining vehicles require custom axles.

The two stripes shown in the middle of each axle designates the type of vehicle that they came from.

This is an early-style Integra intermediate shaft. This one can be used on any B-series swap provided you use the correct axles.

Usually these can be obtained from Honda engine swap specialty shops such as Hasport Performance or Place Racing. However, a few of these transplants are going to require totally custom built pieces, relying on your critical measurements.

Before explaining the measuring process, it's important to take a look at the various intermediate shafts that Honda has to offer, and to understand the differences between them. Who knows, it's possible that you just might have the wrong intermediate shaft and you may not need custom axles after all.

When Honda engineers developed the twin-cam engines, they felt the need to replace the extremely long driver-side axle (found on the Civics for example), with a completely new design. This new driveshaft design incorporated a significantly shorter axle with a dummy shaft fixed to the back of the engine block. Otherwise known as the intermediate shaft or the halfshaft, this dummy shaft transfers power from the transmission to the axle. The design of the opposite side was left virtually unchanged. The intermediate shaft definitely plays a vital role in transferring the power to the pavement.

Engines dealt with in this book that are equipped with intermediate shafts include the B series, ZC, H series, and K series. For the B series, there are several different intermediate shafts available. Right off the bat, you can exclude the automatic units since they won't be used on any of the swaps using manual transmissions. Of the three most common remaining shafts, two of them may be found on the USDM models. Both the '90–'93 Integra SK7 unit and the '94–'01 Integra SR3 unit can be used interchangeably on all B-series transplants. As well as using the corresponding driver-side axle, you must also use the corresponding transmission axle seal. Failure to do so will result in a massive oil leak.

Another common B-series halfshaft is available from the '88–'91 JDM Civic SiR. Although it's compatible with these B-series engines and transmissions, the SH3

intermediate shaft is significantly shorter and will require a custom driver-side axle to be made unless you use the JDM axle that goes with it. You'll also need the axle seal from a '90–'93 Integra.

ZC transmission users will inevitably be using the ZC intermediate shaft found on the '88–'91 JDM Civic Si, not the '86–'89 Integra unit, contrary to popular belief. H-series swappers can choose to use either the '90–'97 Accord SM4 shaft or the '92–'96 Prelude SS0 unit; both are of the manual transmission type. Those transplanting the K series will need the S6M unit from the manual-transmission RSX.

If you've narrowed down your intermediate-shaft selection and found that OEM axles are still not available, measurements may be taken in order to have some made. With your chosen halfshaft in place, measure from the axle seal on the transmission to the outermost portion of the wheel bearing race on the hub. For the side of the vehicle that houses the intermediate shaft, measurements must be taken from the inside lip of the shaft to the wheel bearing race. With the proper numbers, axles can be specially designed to meet your needs. When you're having custom axles made, it's necessary for the builder to know a bit of information about your project. Usually they'll need to know the type of vehicle, the type of engine and transmission, and the type of intermediate shaft you'll be using.

Place Racing has racks of axles in stock for most engine transplants. Give them a call to find out if they have one for your application.

suspension, Prelude and Accord pieces aren't compatible with this chassis. If you don't want to order axles from Hasport Performance or Place Racing, then custom units can be made fairly easily at home. This will require you to use any manual-transmission intermediate shaft from a '92–'96 Prelude along with any '94–'01 Integra axles. You must replace the Integra passenger-side axle's inner joint with one from any '92–'96 Prelude for it to mate up to the H transmission properly. Once fitted together, these axles will perform just as well as any stock unit.

Another solution involves using the intermediate shaft from the '90–'97 Accord and axles from a '90–'93 Integra. Keep in mind that axles using factory components aren't designed to withstand additional amounts of torque and are prone to breakage on Civic-to-H-series engine swaps. Aftermarket units will sometimes be a better way to go, since some companies offer stronger units that sometimes even come with a warranty.

Shifters and Cables Simplified

Of course, you'll need to install a shifter mechanism in order for the car to shift. Unfortunately, the Civic-style shift rods aren't compatible with the Prelude and Accord transmissions. These transmissions were among the first to use the cable-style system that is now present on almost all Hondas and Acuras.

If installed properly, many folks actually prefer these shift cables to the Civic or Integra shift rods. A compatible shifter mechanism for this transmission can be selected from any '90–'97 Accord, as well as any '92–'01 Prelude equipped with a manual transmission. There are several different cables and shifters available in these many Accords and Preludes, so it's important not to mix and match any of these components with one another. You'll find that certain Accord shifters aren't interchangeable with various Prelude cables or other parts.

Once you've selected the assembly of choice, the center console and the carpet around the shifter area must be temporarily removed. Cut the oval-shaped hole into a square that will house the shifter assembly. Make the necessary adjustments with a reciprocating saw to ensure that the shifter plate will sit flat. If the proper cuts aren't made, the shifter will sometimes fail to go into certain gears because of interference with the vehicle's untrimmed sheet metal. Once you cut the

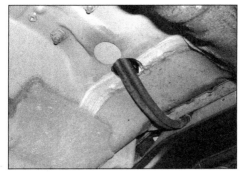

A hole such as this will provide perfect access for the Prelude-style cables to make their way underneath the vehicle toward the gearbox.

hole, bolt the assembly into place with four 8-mm bolts and nylock nuts to ensure that they never rattle loose. Avoid using the floating rubber grommets to mount the assembly, as these will place the shifter up too high, causing it to interfere with the center console.

Next, drill a 2-inch hole about 18 inches in front of the center of the shifter assembly. Use this hole to push the cables underneath the vehicle. Watch out for the airbag wires here, as they'll be right in the way. Be sure to pull them off to the side from inside the car before you do any drilling or cutting in this spot.

Instead of drilling holes and running the cables through the inside, you may opt to use a special mounting box from Place Racing. With this unit, you can mount the shifter cables underneath the vehicle instead of inside. Whichever way you do it, the rest of the installation process is the same. Once they're underneath the car, the cables can now be routed over the rear crossmember and connected to the transmission. Using all of the supplied washers and cotter pins, the cables will attach to the gearbox as stock.

Reattaching the Clutch Line

The clutch line can be reattached in one of two ways. Many folks choose to make a custom clutch line that runs from the clutch master cylinder and to the slave cylinder on the transmission. A piece of -3 steel braided line may be used if you choose to go this route, as the pressure rating is plenty suitable for a hydraulic clutch line such as this. You'll

When reinstalling the shifter cables on the transmission, be sure to use all the factory-supplied washers and cotter pins.

need some 90-degree hose ends and 10-mm to -3 adapters to thread the line into both clutch cylinder components. For those who don't want to spend the extra cash on a fancy AN braided line, the original clutch line can be bent and persuaded to reattach itself to the Prelude slave cylinder. Whichever way you choose, you'll need to bleed the air pockets out of the hydraulic clutch system when you're finished.

The next thing you'll need is a throttle cable. A throttle cable from any '97–'01 Prelude will snap right into place. If crawling underneath the dash isn't something you look forward to, then a special bracket is available from Place Racing that will allow you to reuse your original throttle cable on the H engine.

When using the Prelude master cylinder, a metric-to-AN fitting may be threaded into place in order to attach a custom steel braided hydraulic clutch line.

Finishing Off the Fuel System

As far as the fuel system, things aren't going to simply reattach as easily as in previous swaps. The fuel injection feed hose is going to pose the biggest problem of all. In many cases, the inlet on the Prelude fuel rail is located on the opposite side as it was in the Civic. This makes attaching the original Civic line to the Prelude fuel rail rather difficult. An easy fix is to install the '92–'96 USDM fuel rail, which has the inlet on the opposite of the JDM unit. After you install the new fuel rail, the original Civic fuel injection feed line will hook right up. To wrap up the fuel system, you'll need a reducer to go between the smaller Civic fuel return line and the larger outlet on the Prelude fuel pressure regulator. As usual, use fuel-injection

hose clamps and hose to avoid any possible leaks down the road caused by inferior, low-pressure line.

Cooling System Procedure

The cooling system isn't totally straightforward either, but it's a little bit easier than the fuel system. The original radiator of choice can be bolted back into position, but the original cooling fan will no longer fit. Since it will interfere with the clutch slave cylinder, a slimmer fan must be installed either on the front or the back of the radiator. You can wire up most of these fans to push or pull air, depending on which side of the radiator it is mounted. Be sure to hook up the two wires correctly to the original fan's connector on the vehicle harness.

Notice how the smaller radiator fan sits perfectly in the space between the bumper and the radiator. Simply connect these two wires to the appropriate locations the same way the old fan plug was connected.

This Del Sol DOHC VTEC twin-core radiator is a must-have when installing the H-series engine. The original Civic unit is just too small and isn't up to the task.

After you've installed the radiator, you can get some radiator hoses from '92–'96 Prelude. You'll need to trim the top hose somewhat, but they can both be attached to the Prelude engine with the original clamps. Unless you use the twin-core Del Sol DOHC VTEC or Civic Si radiators with the larger inlets and outlets, you'll need to use adjustable hose clamps to attach the two hoses to the radiator. The adjustable clamps are necessary to avoid leakage because of the difference in size between the radiator inlet and outlet and the hoses.

The last modification to the cooling system involves the heater hose that was attached earlier to the water inlet on the firewall. The other end of this hose needs to be connected to the rear water pipe outlet on the back of the engine block. Use an adjustable hose clamp for this connection as well. With all of the hoses in place, the radiator can be filled up with the proper mixture of Honda coolant and water.

A/C Retention

For those of you who wish to retain your air conditioning, be forewarned: it is definitely a tight fit. By using a special A/C bracket from HCP Engineering, Place Racing, or Hasport Performance, you can reuse the Civic air compressor with the H-series engine. You'll need to make some special brackets for the condenser since the factory ones will need to be cut off for clearance. A slim fan like the one being used for the cooling system will need to be installed on the front side of the condenser in lieu of the original unit. Fortunately, the original Civic A/C hoses can be retained, although they're a tight fit. Retaining the A/C on this transplant will prove to be a labor-intensive job; it can sometimes take up half of the entire project's allotted time.

Keeping the Power Steering

In addition to the A/C, you might also want to keep the power steering. It can work if you use a '92–'96 Prelude power steering pump and bracket and a bit of work. Since the EJ/EM power-steering hose is incompatible with the H-series pump, you'll need to cut the line and replace it with the hose end from either a Prelude or an Accord. The new hose end must then be brazed or TIG welded to the old hose, which can then be bolted to the pump. Since original Prelude components are being used, so can the power-steering-pump belt, which will also drive the alternator. To avoid contact between the power-steering pump and the headlight, do some minor trimming to the plastic housing of the fixture. You'll need this clearance for when the engine rocks.

Necessary Upgrades

Although the H-series swap will add 185 pounds to the front of the vehicle, the adverse effects will prove to be a bigger deal for some folks than others. For those who generally spend their weekends autocrossing their Civics or tearing up the corners of a road coarse, the heavier engine is probably not for you. However, for those enthusiasts whose need for straight-line acceleration takes precedence over the car's overall ability to handle and maneuver, the H series will probably suit you just fine.

For those who do choose the Prelude route, the extra weight is definitely going to take its toll on the suspension system. Special engine swap springs from Progress Suspension will be necessary to keep the ride height at an acceptable level, while still giving the vehicle a lowered

Although it's time consuming to say the least, few swaps provide the enjoyment that an H series does. Once installed properly, you'll find that it fits snugly right under the Civic hood.

stance for optimal handling. Standard Civic lowering springs will sag down significantly, since they're not designed for the additional weight up front.

A brake upgrade in the form of four-wheel discs is recommended for this swap in the name of safety. With the addition of '94–'01 Integra or '99–'00 Civic Si front and rear disc brakes, the Civic's stopping power will be more than adequate for the likes of the more powerful engine. Further brake upgrades can be found in the form of heavy-duty brake pads and slotted rotors from Brembo or AEM.

A Word for the H Series

This transplant into the '96–'00 Civic adds so much weight and isn't the most reliable, so it's a wonder why anyone would ever choose such a project. If you've ever been behind the wheel of one of these cars, then you wouldn't wonder anymore. With acceleration comparable to that of a turbocharged Integra or Civic, the '96–'00-Civic-to-Prelude swap will give most V-8s a run for their money. Moreover, with just a few minor bolt-on modifications, your H series can make over 200 horsepower at the wheels. It would take some serious cash to generate this type of power from any run-of-the-mill B series. In addition, the part that most enthusiasts like best is that the Prelude engine is cheaper than the B series to begin with.

Notice how the rear brakes were converted with components from a late-model Integra. You can see the larger calipers on the back portion of the rotor.

2001 TO 2003 CIVIC

The introduction of the '01 Honda Civic, otherwise known as the ES/EM chassis, unveiled many new changes for this middle-aged compact car. A reverse-oriented engine in the Si, as well as the standard implementation of the Prelude-style shifter cables are but two of the major changes. In addition to the obvious changes made to the body, it is the mechanical differences like these that true Honda freaks notice right off the bat. It's been manufactured for only three years so far.

Civic Offerings

Body styles range from coupe to hatchback to sedan, and several trim levels are available. Among the most popular and most luxurious of the Civics, the EX is still available in both the two-door and four-door form. It features the most powerful engine in the new lineup (excluding the Si), the D17A2 SOHC VTEC engine.

Speaking of the Civic Si (EP chassis), it has only been produced in '02–'03 so far, and it is the most powerful Civic available. It features the base-model RSX engine, the K20A3, which puts out a respectable 160 horsepower. The LX and DX versions are both a step down, each available with the less-powerful D17A1 engine. The HX is yet another Civic model, equipped with the D17A6. Unlike the LX and DX D17A1, the HX D17A6 features VTEC, although it isn't as powerful as the EX Civic. A GX model is still available, and once again, it is equipped with the natural-gas engine. It's one of the cleanest burning vehicles in the United States, and most of its owners probably aren't going to be interested in

This isn't the most common swap just yet, but soon enough you'll see RSX engines being transplanted into Civics of all sorts. (Photo Courtesy of Hasport Performance)

Although it has only been out for a couple of years, several engine swaps have already been completed on this generation Civic, and several more are in the works.

a larger-displacement, gasoline-burning, high-performance engine swap.

Braking and Handling Upgrades

Thanks to another close relationship between the Civic and its older brother, several components can be shared between the Civic and RSX chassis. Both the suspension and braking systems are completely compatible with the Civic; so look no further than genuine Honda parts when searching for upgrades. Further enhancements can be made with the help of Progress Suspension; which has several high-performance suspension components already available for the ES/EM Civic. Handling concerns will be the least of your post-engine-swap worries.

K-Series Engine Swap

The seventh generation of Honda Civics yielded some surprising traits, namely, its resemblance to the RSX (new Integra) engine bay. Much as the '92–'95 Civics shared many characteristics with the '94–'01 Integras, the little-brother Civic and big-brother Integra tradition continues.

The complexity of the K-series transplant is mostly attributed to the fact that very few people understand the

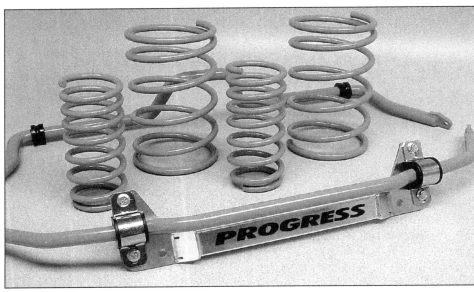

These engine-swap designed springs and sway bars are necessary for any Civic with an RSX engine transplant. Progress Suspension has had these in the works for some time now. (Photo Courtesy of Progress Suspension)

You want this one. The top of the line RSX engine will provide you with the most bolt-in horsepower possible thanks to the Hasport Performance mount kit.

'02-'03 Civic Si

The RSX drivetrain bolts perfectly into the '02–'03 Civic Si. No need for any special aftermarket engine-swap mounts here. Even the RSX engine harness can be reused. The Si has 160 hp, while the RSX Type S has 220 hp.

The top of the line as far as seventh-generation Civics are concerned, the Civic Si is one of the most performance-oriented Hondas available.

The RSX base engine can also be found in the Civic Si model. These may be swapped into the Civic chassis as well, but they're seriously lacking in the horsepower department compared to the Type S.

inner workings of the K-series engines, let alone the '01–'03 Civic. At this time, it's recommended that a swap like this be left to the professionals. The folks at Hasport Performance offer the one motor mount kit available for the '01–'03 Civic-to-K-series transplant. Professionals such as the people at Hasport Performance can do this swap in only a day's work.

K-Series Engines and Transmissions

Since they were introduced just a couple of years ago, very few K-series engines even exist on the roads, not to mention in the junkyards. Perhaps the most popular and abundant of the

bunch, the K20A3 is the base model engine of the RSX. Found in the '02–'03 USDM RSX and Civic Si, these engines are the most affordable K series.

A similar JDM version of this engine is also available; these are stamped as simply the K20A, without any numerals following. The top-of-the-line K-series engine can be found in the USDM RSX Type S K20A2. Pushing 220 horsepower, these are the most powerful four-cylinder engines produced by Honda for the U.S. market. A similar JDM version called the K20A can found in the Civic and Integra Type R. All of these engines are similar in external dimensions and weights, mak-

ing the transplant processes virtually identical for all. All K-series swaps into these chassis will require you to use a K-series transmission.

ECUs and Wiring for the K Series

With the most advanced OBD system to date from Honda, very few choices exist when selecting the proper computer for this transplant. Unlike other swaps where you'll be able to exchange vehicle harness plugs to adapt a different ECU, this won't be possible with the ES/EM-chassis Civic. It's important to use the ECU designated for use with the K20A engine at all times. Even using an ECU from the

Hasport Performance

The Honda engine-swap-mount business has become crowded in the last couple of years with many up-and-coming companies now manufacturing mounts. In spite of increased competition, proprietor Brian Gillespie has been able to maintain Hasport Performance's position on top of the heap due to several competitive advantages.

If you're trying to engineer the best engine-swap components, it's very helpful to look and see how Honda did it in the first place. Hasport Performance is in a unique position because of their close association with Honda Auto Salvage. This gives them access to any Honda automobiles that they need, including the Japanese counterparts. This allows the folks at Hasport Performance to see the bigger picture when making decisions about ideal engine placement and axle angle. Another advantage to this is the ability to document subtle differences in trim levels and models to ensure the engine mount kits and instructions are correct for each customer's car.

Hasport Performance makes full use of current technology to design and produce the best product possible. Mounts are designed and modeled in house using the industry standard CAD/CAM software, Pro/ENGINEER. Using the integrated design analysis software Pro/MECHANICA, the models can then be tested for structural design optimization. Pro/ENGINEER Wildfire is then used for CNC programming so that all tool paths can be optimized, saving valuable machining time.

Hasport's mounts are CNC machined in-house using one of three HAAS VF-4 Vertical Machining Centers. To increase productivity, automatic pallet changers are incorporated so that non-machining tasks, such as fixture setups and part changeovers, can be performed off-line on one pallet while parts are being machined on the other pallet. This helps Hasport keep back-orders to a minimum and offer the best price possible. Custom tooling is also utilized to further shorten mill time. As productivity increases, prices for Hasport Performance products will continue to drop, says Gillespie.

The mounts are machined from solid 6061-T6 aluminum at tolerances of less than 1/1000 of an inch. Because of heat and warping, the welded steel mounts of their competitors can't duplicate these tolerances. With aluminum's unique combination of strength and light weight, Hasport mounts are pound for pound much stronger than steel, and more attractive too. This is why they confidently offer a lifetime warranty on their engine mounts. Since switching from welded assemblies to machined billet, Hasport Performance's warranty rate has been reduced to less than .002 percent.

The polyurethane bushings used in Hasport Performance mounts are produced by one of the largest manufacturers of polyurethane products for automotive and motorcycle use. Over the years, the bushing's design and composition has been upgraded to eliminate problems associated with the extreme abuse associated with high-performance driving. Their current bushing has twice the tensile strength of their original bushing, and therefore they carry the same lifetime warranty as the billet mounts they're used in.

Hasport Performance is now in the process of testing a revolutionary new bushing design and compound. It's a micro cellular hybrid product that will give a factory-like feel yet still have the durability you expect from a Hasport Performance product. This new technology will allow them complete control of the durometer without the decreased service life associated with softer materials. The look will change too. The new bushing will be locked into place from the inside, which will result in a smoother, more streamlined look. These advancements should further separate Hasport mounts from the pack.

While Hasport Performance knows their product is second to none, they recognize that customer service is a necessary ingredient in any business as well. Their sales staff is intimately familiar with their products, and with Honda engine swaps in general. Their mission is to help the customer make the right decision before the sale and then be able to finish the project once it is started. In addition to engine mounts, Hasport Performance sells other specialty products associated with the swap such as shift linkages, axles, ECUs, and plug-and-play wiring solutions.

These guys recognize that an engine swap is usually just one modification in a long list for most Hondaphiles. That's why Hasport Performance has recently teamed up with Jackson Racing, Hondata, Rev Hard Turbo, Progress Suspension, and AEM in order to bring swap-specific products from these manufacturers to the Honda performance enthusiast.

Hasport Performance prides itself on being one of the driving forces in the popularity of the Honda engine swap. Through sponsorship and participation, they maintain a high profile at trade shows, in import racing, and television. Do a quick review of all the how-to articles published on Honda engine swaps, and you'll find Hasport Performance is the most-often used source for products and information. Hasport Performance will continue to lead the industry with products for Honda's new K-series engines as well.

RSX isn't as simple as it has been with older ECUs. Honda-implemented anti-theft features require the new computer to be flashed in order to accept the Civic's key code. You can get this done by visiting a local Honda or Acura dealership post swap. Be sure to bring all of the proper VIN information for the RSX engine.

Folks with JDM engines will have to find another option. According to Brian Gillespie of Hasport Performance, using an RSX engine wiring harness is the smartest way to go with this transplant. Although the Civic harness could be modified and reused, the RSX harness has most of the painstaking wiring already completed for you. All that needs to be done are some basic modifications and to add some wiring provisions for a few additional components. Before installing the wiring harness, the two heavy-gauge wires that attach to the fuse box must be moved to the other side of the engine bay to reach the Civic's fuse box. The positive and negative cables will then route to the side of the engine where the RSX's battery would rest.

When using the RSX engine harness, you'll find that the Civic vehicle harness is lacking three items. For the engine to function properly, you must appropriately wire the air/fuel-ratio sensor, along with its relay, the air/fuel-ratio display, and the radiator fan switch. HX model Civics can skip over

The ECT (engine coolant temperature) switch wiring must be lengthened on the Civic harness if you're installing the Civic Si radiator. This is the new location of the plug on the new radiator. (Photo Courtesy of Hasport Performance)

the fan switch, however. You must modify it since it's located on the Civic's engine, as opposed to being found on the RSX's radiator.

Final wiring modifications include swapping the RSX coolant temperature switch for the one found on the original Civic engine. This must be swapped out due to its incompatibility with the Civic's multiplex control unit. If all of this sounds too confusing, or you simply just don't have time to mess with the wiring, Hasport Performance offers complete plug-and-play wiring conversions for this swap.

You need to attach the Hasport Performance side mounting brackets to the frame before the engine goes in. It's much easier to do it like this. (Photo Courtesy of Hasport Performance)

Installing the K Series

After you've properly modified the engine harness and installed it onto the new engine, the appropriate Hasport Performance mounts can be put into place. The Hasport Performance ESK1 kit is a five-piece kit consisting of two steel brackets and three CNC-machined billet-aluminum motor mounts. You'll want to start things off by bolting the left-side engine bracket into place with the original hardware.

If the vehicle is equipped with anti-lock brakes (ABS), you'll need to shorten the ABS bracket so that it won't interfere with the mount once installed.

The wiring is being positioned onto the engine and fished into the vehicle in this photo. A custom harness from Hasport Performance was used because the electrical system is fairly complex on this swap. (Photo Courtesy of Hasport Performance)

Here's a look at the '02–'03 Civic Si radiator installed into this Civic. If you look toward the top, you can see where it had interfered with the hood latch. (Photo Courtesy of Hasport Performance)

This Hasport Performance billet driver-side mount is about as strong as you can get. With the two side mounts in place, the engine will sit fine. (Photo Courtesy of Hasport Performance)

Once you've shortened the ABS bracket, drill a hole through it and the left-side engine bracket. Next, attach the right-side engine bracket to the frame rail with the provided bolts and spacers.

The factory Civic radiator inlet and outlet ports are in awkward positions relative to the new engine, so they won't work too well. The radiator, cooling fan, and condenser fan all need to come from a Civic Si, so you'll need to make some adjustments to the core support of the vehicle in order for them to fit. Before the new radiator can be installed, you

have to cut out the factory hood latch support and replace it with a modified unit from Hasport Performance. Once you install the modified unit, the Civic Si parts will fit perfectly.

If you need extra clearance for the battery, you may need a slimmer cooling fan in lieu of the Civic Si unit. Lower the engine into the bay at an angle, attach the passenger-side engine mount first, and then lower the transmission side of the engine. Remember, these engines are reverse oriented compared to everything else in this book, so the installation

process might seem a little different. Next, you need to install the left-side engine mount. With that, remove the engine hoist, as the two mounts will support the engine just fine.

The new Civics and RSXs have a subframe mounted beneath the engine, which is different from any of the other Hondas and Acuras in this book. You'll need a subframe from the RSX or the '02–'03 Civic Si for the RSX engine to sit properly in the Civic.

Once you've bolted the subframe into position, combine a Hasport Performance rear mount and an RSX bracket to install it. You can also install an RSX front mount. With that, the engine is properly set into place.

Unfortunately, since the seventh-generation Civic chassis isn't identical to the RSX chassis, a couple of clearance issues will arise during the swap process. Even after you trim the core support for the new radiator, the base-model RSX engine will still have trouble fitting. The K20A3 has a different intake manifold, so it won't clear the core support. Luckily for you, the engine that does fit is the more powerful of the two, the USDM K20A2. Other than those minor clearance issues, the K-series engine sits underneath the hood quite nicely, providing a factory-looking fit.

Suspension and Axle Options

You have a couple of options concerning suspension components for the Civic. Although the Civic suspension can be reused, the RSX front knuckles,

This subframe is from the RSX and is going to be installed on the Civic chassis. Attach the sway bar and rear mount before bolting it into place. (Photo Courtesy of Hasport Performance)

Once the subframe is raised into position, the rear bracket of the K-series can be attached to the mount and bolted down. (Photo Courtesy of Hasport Performance)

These RSX axles slide right into place along with the RSX front knuckles and braking components. (Photo Courtesy of Hasport Performance)

Understeer Solutions

Understeer is most common to front-wheel-drive vehicles and occurs when the front tires have a greater slip angle than the rear tires. In other words, even when you keep turning the steering wheel (even significantly farther), the car won't turn any sharper. The result is that the car continues in more of a straight line than in a harsh turn. Stock production cars have built-in understeer already; therefore installing a heavier engine only makes matters worse. The heavier the engine and the farther forward it is, the worse the effects of understeer will be. Installing a heavier B- and H-series engine into a Civic is a perfect testament to these effects. Several common solutions for fighting the understeer battle are listed below:

1. Decrease front tire pressure and increase rear tire pressure.
2. Go with a wider front tire section and a smaller rear tire section.
3. Add more negative front wheel camber and more positive rear wheel camber.
4. Add more front wheel toe out and more rear wheel toe in.
5. Add more positive front wheel caster.
6. Go with softer front springs and stiffer rear springs.
7. Go with a softer front sway bar and stiffer rear sway bar.
8. Move more weight to the rear (consider this prior to any vehicle weight reduction).

ABS sensors for this conversion to work. If you change over all this, you can also use a set of unmodified RSX axles. This is by far the best route to take, but it's also the most expensive.

If changing over the suspension isn't an option for you, RSX axles can still be used, but they'll have to be modified. Because of differences in the axle-shaft diameters between the Civic and the RSX, custom axles will need to be made. Rather than purchasing a pair of custom-built axles, you can use outer joints from a '94–'01 Integra mated to the RSX shafts to achieve a perfect fit.

Another even less expensive axle solution is to use unmodified '02–'03 Civic Si axles. You can simply slide them into place. Of course, you won't have the stout RSX suspension and brakes, but it will be significantly less expensive. In all axle scenarios mentioned here, always use the K-series intermediate shaft: it's the only one that will work.

Shifter Instructions

With a manual-transmission Civic, the original shifter mechanism can be reused on the RSX transmission. You can reinstall it with the original hardware, and you'll find it's almost a perfect fit.

Due to slight differences in the Civic and RSX shifters, you'll need to make some minor modifications to the Civic unit in order to engage reverse.

you should be able to freely move the shifter into reverse.

Hooking up the Civic hydraulic clutch system to the RSX transmission isn't too difficult. Use an RSX clutch fluid line to connect the vehicle's clutch master cylinder to the slave cylinder, and then fabricate a small bracket to hold it in place. Next, you'll need to add a throttle cable and a brake booster vacuum line from the RSX to finish off the

pedal assembly. Don't forget to bleed the clutch line once you're finished.

Easy Work of the Fluid Systems

Reattaching all the hoses and lines for the fluid systems isn't too tricky. Start with the fuel system. The main fuel line and purge line on the vehicle must be bent gently so they can attach to the RSX lines. Be careful not to kink the lines, as it is very labor-intensive to replace them. If your RSX engine doesn't come with the lines, you may need to purchase them separately.

Install a radiator from a Civic Si, and then all that is necessary to finish the cooling system is to add upper and lower radiator hoses, also from an '02–'03 Civic Si.

A/C and Power Steering How To

You'll need some more components from the Civic Si to keep things cool inside the vehicle. Use an RSX A/C compressor and an RSX line to run from the compressor to the condenser. A custom A/C line must be made to run from the condenser to the firewall to complete the system.

Reinstalling power steering isn't much trickier. Use the RSX power-steering pump, and swap out the Civic power-

These Civic shifter cables attached to the K-series gearbox in factory form. All of the washers and cotter pins were reused for a perfect fit. (Photo Courtesy of Hasport Performance)

steering lines for RSX units. You'll be lacking provisions to mount the new lines on the passenger side of the steering rack, but you can use some heavy-duty zip ties to fasten the hoses down securely. A '92–'95 Civic power-steering reservoir will fit perfectly in front of the fuse box, but a custom bracket must be made to hold it down tight.

Finish off the power-steering system by modifying an RSX return line to reach the new reservoir.

You'll need to create some clearance under the hood by trimming away some of the underhood skeleton where necessary. This extra clearance between the pulley and the hood helps avoid body damage.

Sharing With the RSX

Compatibility between the Civic and the RSX means that fortunate Civic owners cannot only swap over front suspension and brakes, they can also perform a rear disc-brake conversion. Replacing the rear drum-brake setup of the '01–'03 Civic with the stout rear discs of the RSX will greatly reduce stopping distance, resulting in a safer car post engine swap. Progress Suspension specially designed lowering springs to significantly improve the handling of the engine-swap vehicle as well. By combining these springs with Progress Suspension's alignment adjustment kit and rear anti-sway bar, it is possible for the vehicle to handle better than it did before, despite the extra weight of the new engine.

Wrapping Up the K Series

There certainly aren't a lot of options right now for seventh-generation Civic owners who are determined to go fast. Those available are definitely among the most expensive in this book. Nevertheless, thanks to the folks at Hasport Performance, engine swaps for these vehicles are now possible. Although the initial cost of the K-series swap is extremely high, it is a much more reliable means of a power gain than a turbocharger or nitrous oxide kit on these less-than-stout SOHC engines. Providing a factory fit and complete emissions-legal status in most states, the K-series swap should definitely be considered by all power-hungry '01–'03 Civic owners.

You can solve your power-steering reservoir problem by using one from an older Civic. It can be slid into place and will work fine. (Photo Courtesy of Hasport Performance)

The power-steering line from the RSX is used to attach to the RSX power-steering pump. It bolts into place with the right components. (Photo Courtesy of Hasport Performance)

This is a rather rare swap at this time; you're more likely to see them at the racetrack or on exhibition somewhere. As the prices of the engines come down, more swaps will surely hit the road.

1998 TO 1991 PRELUDE

When the '88–'91 Prelude was introduced, so was a substantially more powerful and sportier version of Honda's top-of-the-line sports car. The new third-generation Prelude was equipped with its first DOHC engine and could actually back up its aggressive looks with 140 horsepower underneath its hood. The Prelude has always been more expensive than the Civic and far sportier than the Accord, and the BA chassis Prelude had its loyal following back then just as the newer versions do now. Although not the most popular vehicle of its time, the Prelude has always been one of the most fun to drive.

Prelude Offerings

The Prelude was only available in two different trim levels; the S is the entry-level model, while the Si sports the more powerful engine and leather-clad interior. Different versions of the Si became available after 1989, ranging from the 2.0Si, the 2.1Si, the Si-ABS, and the Si-4WS models. Regardless of which trim level you choose, all Preludes produced during this time are only available in two-door coupe form, with a rather small back seat. Available S model engines include the B20A3 engine, which was offered only with a carburetor. These were only sold during the

model years '88–'90. The Si model came with the B20A5 engine.

A more powerful B21A1 engine can be found in select '90–'91 Si models. All engines excluding the B20A3 use fuel injection systems in lieu of carburetors. As we discussed earlier, these engines are B series only by name, and are far different from the B-series engines of the Civics and Integras that we've come know and swap so often.

OBD Issues

Most of the '88–'91 Preludes are OBD 0 (very few '91 models were OBD I); so a wiring conversion will be essential

An uncommon sight to say the least, you won't find many H22A swaps into the '88-'91 Prelude chassis.

This is one of the most rare engine swap vehicles; very few '88-'91 Prelude owners actually go this route. Preludes sport a DOHC engine to begin with, so they often try to upgrade their engines instead.

Place Racing

One of the largest engine-swap parts manufacturers, Place Racing opened for business in May 1992. The import scene as we know it, revolving around engine swaps and enormous turbocharged horsepower figures, hadn't even been thought of yet. Gil Garcia opened shop in an old 5,000-square-foot industrial building. The place was huge according to the guys who worked there. When Place Racing first started, they couldn't figure out what to call the new business, so they tossed around a few ideas. Eventually they began just calling it "the place" and the name stuck. Later on, their decal designer was making up some stickers for the guys and he added on the word racing after the name. Garcia and the guys liked it so much that they started calling themselves Place Racing. When the sales people answered the phone, they said "Place Racing," and it stuck. When incorporating in 1997, they kept the title and became Place Racing Inc. Sometimes they shortened it to just PRI.

You might consider Gil Garcia a pioneer of the Honda engine-swap craze. Garcia has definitely paved the way for the Honda engine-swap industry, being one of the first to swap a ZC and B-series engine. At the start of Place Racing however, Garcia and company didn't have the slightest idea they'd someday be manufacturing components instead of just reselling them. In fact, they really didn't offer much in the way of engine parts at all. Before the Place Racing engineering department even started thinking about manufacturing any type of mount kit, they started manufacturing custom intake tubes. They would cut and weld these intakes and hand paint them to the customer's specs. Long before anyone even thought of "cold air intake," Place Racing was building the intakes for the Civic, according to Garcia.

Through their first years in business, they outgrew two facilities before settling down at the current location in Azusa, California. Even at the new facility, they have had to expand from one building into three. The manufacturing and assembly plant uses two buildings, and the sales and shipping departments occupy the third. Place Racing has gone from one small welding machine to owning and operating some of the latest in CNC-controlled machinery. Place Racing uses two HAAS CNC Vertical Machining Centers, as well as CNC bending, punching, and folding machinery. Its manufacturing department also consists of computer-controlled cutting and mandrel-bending equipment. It manufacture more than 99 percent of its own parts starting from the raw material all the way through the finished product. Since the beginning, Place Racing has stood behind every product that it manufactures. Itsalways offered an unconditional lifetime warranty on all of its products.

Unlike the all too common fly-by-night import companies, Place Racing strives to manufacture the best parts that they possibly can and backs each with its unconditional warranty. Each member of the PRI team, from the engineering department to manufacturing to sales and shipping, takes pride in the products that bear the Place Racing name. According to Garcia, PRI is and always will be the nation's largest producer of engine-swap components for the import market.

Place Racing Honda engine-swap products include:

'84–'87 Civic-to-B-series mount kit
'86–'89 Integra-to-B-series mount kit
'86–'89 Accord-to-B-series mount kit
'88–'91 Prelude-to-H-series mount kit
'88–'91 Civic-to-B-series mount kit
'88–'91 Civic-to-B-series shift linkage
'88–'91 Civic-to-H-series mount kit
'92–'95 Civic and '94–'01 Integra-to-H-series mount kit
'96–'00 Civic-to-H-series mount kit

'90–'93 Integra auto-to-manual conversion mounts for cable B-series transmissions
'92–'95 Civic auto-to-manual conversion mounts for cable B-series transmissions
'92–'95 Civic auto-to-manual conversion mounts for hydraulic B-series transmissions
'94–'01 Integra auto-to-manual conversion mounts for hydraulic B-series transmissions
'88–'91 Civic cable-to-hydraulic transmission adapter kit for B series
'88–'91 Civic cable-to-hydraulic transmission adapter kit for H series
'90–'93 Integra cable-to-hydraulic transmission adapter kit for H series
'92–'95 Civic and '94–'01 Integra hydraulic-to-cable transmission adapter kit for B series

Custom crossmembers and axle solutions for most transplants.
Throttle cable, A/C, power steering, and specialized brackets for most transplants.
Customized engine wiring harnesses and OBD adapter harnesses for most transplants.

USDM OBD II computers due to their variety of necessary emissions sensors. Modifying the vehicle for a USDM OBD II ECU would cost you days of unnecessary wiring. Bottom line: when going OBD II, stick with the JDM boxes.

In any case, using any OBD computer will require the vehicle harness to be modified. Plugging an OBD ECU into the Prelude's car harness isn't going to happen. Therefore, you need the proper ECU connectors from a junkyard vehicle. Cut off some OBD I plugs from any '92-'95 Honda several inches away from the ECU. You can find OBD II pin any '96-'98 Honda; cut them off in the same manner. Although OBD II by name, '99-'01 connectors from Civics and Integras won't work.

After you've acquired the proper connectors, the wires must be individually soldered into place on the Prelude underdash harness. Replace the ECU electrical connectors one by one, so the new plugs will allow the new computer to plug right into place. Since you won't be able to read engine trouble codes in Morse code fashion on the cover plate of the new ECU, you'll have to view them in a new manner. Add the appropriate wiring so the malfunction codes can be viewed on the gauge cluster. Of course, you'll need the service manuals for the Prelude and the new ECU to complete the wiring.

H-Series Engine Swap

Certainly one of the most unlikely of engine swaps that you'll read about in this book, an '88–'91 Prelude with H-series power is, without a doubt, a monster in disguise. The third-generation Prelude has a major advantage in that it weighs significantly less than the newer fifth-generation Preludes, tipping the scales at around 2,600 pounds. Swapping in the most powerful H engine, the H22A will definitely wake up this tired old chassis. Place Racing made this swap a straight bolt-in affair, making it an intermediate level swap project. Several electrical issues need to be addressed, so the novice engine swapper might consider leaving the wiring harness to a professional. This is the most complex part of the transplant.

This B20A5 engine is hopeless as far as horsepower is concerned. These engines feature a composite head gasket as opposed to the standard metal version, so they can't withstand much horsepower.

This JDM ECU is smaller than the USDM version, which makes it an easy find. This particular computer is already equipped with the plugs necessary for the OBD conversion.

Not including the electrical modifications, this swap can generally be completed by a pro in a day's work.

H-Series Engines and Transmissions

Since third-generation Preludes are OBD 0, your first instinct would be to select an OBD 0 donor engine. You can quickly rule out that option, since H-series engines aren't available in anything older than OBD I. With that being said, pretty much all of the '92–'96 H series engines are fair game, as are the corresponding transmissions and computers (excluding the '96 ECU). Engines from '97–'01 will also work, but their ATTS transmissions and USDM ECUs are useless for this particular swap. Although the ATTS is a work of genius from the engineers at Honda, unfortunately these units won't bolt into this chassis without some major fabrication. Even if it could fit, the wiring involved in the ATTS computer would be way too time consuming. As far as the USDM '97–'01 ECUs are concerned, the built-in anti-theft device makes these computers the least desirable for this swap. If you still want a '97–'01 engine, keep in mind that the transmission and ECU must be replaced with other units. Most professionals recommend setting

your sights on one of the earlier engines.

With that said, let's briefly cover the engines that are compatible with the '88–'91 chassis. The H22A is the most popular swap into most any Honda or Acura vehicle, so you just can't go wrong with it. Offering more power and torque than any B series, the Prelude VTEC engine is the number-one engine swap option for the BA chassis Prelude.

The H22A1 can be located from most wrecking yards out of the '93–'96 USDM Prelude Si-VTEC. The slightly more powerful JDM Prelude VTEC engine of those same years, in addition to 1992, is known as the H22A, without the number code following it. Other VTEC offerings compatible with this chassis can be located in the '94–'97 JDM Accord SiR. Non-VTEC engines known as the H23A1 can be found under the hood of the same Preludes just mentioned, but sporting the Si badge instead of the VTEC emblem.

Several other EDM and OBD II H-series engines are available, but for this particular chassis, they're generally more expensive than the car itself. As with any of the Prelude engine swaps, the transmission of choice is the Prelude VTEC unit that is usually mated to the H22As straight from the junkyard. Due

to its shorter gear ratios, even those purchasing an H23A sometimes select it for its race-inspired feel. These can be acquired from '93–'96 USDM or JDM Prelude Si-VTECs and select JDM Accords. For those in search of the utmost in performance, rare limited-slip versions of this gearbox exist in Japan and can be found every so often in engine yards here in the United States.

ECUs and Wiring For the H Series

Electrically speaking, quite a bit of wiring is involved in the H-series transplant for this car. Since we're talking about an OBD 0 chassis, an OBD conversion is absolutely necessary to use the correct Prelude computer. For the sake of simplicity and performance, an OBD I conversion is recommended as opposed to OBD II. For this to work, you'll need to modify the ECU connectors on the vehicle's underdash harness to adapt to the newer-style connectors.

No plug-in adapter harnesses are available for this conversion, but you can find the plugs you need in any OBD I Honda vehicle and solder them into place. You can figure out the proper connections by consulting the schematics of the '88–'91 Prelude and '92–'95 Prelude service manuals. Once you get that done, you'll be able to connect any OBD I computer. Computers you can choose from include the P13 ECU as well as any OBD I B-series VTEC ECU for all VTEC transplants and the P14 for all non-VTEC transplants. Both of these computers can be found in USDM and

The H22A1 Prelude VTEC engine is the most common H swap for any chassis. In this photo, you can get a good view of the secondary-type intake system of the Prelude intake manifold.

Since the cylinder position sensor is driven off of the exhaust camshaft on the B20A5, wires will need to be switched around inside the harness to adapt it to the Prelude distributor.

JDM '92–'95 Preludes. Remember, even though all of the '92–'96 Preludes are similar in body style and drivetrain, the '96 OBD II ECU is incompatible. If you still want OBD II despite these warnings, stick with a JDM unit that does not require the additional emissions components. Keep in mind, though, that you'll still have a lengthy wiring conversion ahead of you.

Before you install the original Prelude's wiring harness onto the new H-series engine, you need to swap out the injector plugs for late-model connectors. Once you've soldered them into place, attach the harness with the fuel-injector clips first. This will allow you to distinguish the connectors that need their wires lengthened to reach their respective sensors.

In addition to adjusting the length of the wiring harness, you'll need to wire in an intake air bypass sensor and

an EGR valve when using the P13 and P14 ECUs. You'll also need to make some VTEC connections when using all H-series VTEC engines, including the VTEC pressure switch and the VTEC solenoid. Note that some late-model JDM H22As do not require a VTEC pressure switch. This is only an option if you use the ECU designed for that respective engine. If you're using a different ECU, the entire VTEC solenoid unit must be swapped out for one with provisions for a pressure switch. As far as the distributor wiring is concerned, if you install an H engine with an internal ignition coil, the external coil setup of the Prelude must be converted. Many H22A and H23A engines use external-coil distributors, so this job should be fairly plug and play.

To finish up the harness, you'll need to graft in the alternator portion of the H-series harness, add additional wiring

for a knock sensor, and swap out the oil pressure sensor connector and reverse light connector for prelude connectors.

Installing the H Series

Unless you want to bust out the welder and some fabricating tools, Place Racing is your only option when it comes to H-series engine mounts for this Prelude chassis. The Place Racing kit consists of two high-quality mild steel engine mounts and a special rear adapter bracket, making it a straight bolt-in deal.

With so few components involved, the installation process will be anything but confusing. Extra steps are necessary for those with the earlier '88–'89 models. Enthusiasts with the earlier chassis will need to acquire the driver-side frame bracket from the '90–'91 Prelude and bolt it onto the frame rail. In contrast to most Honda and Acura frame rail pocket mounts, this isn't a welded piece, so it can easily be bolted into place. Once the new bracket is in position, the new Place Racing engine mount can be attached to the body. Place Racing also supplies the bolt-on right-side engine mount for all four model years of this Prelude.

The rear engine bracket of choice comes from the '92–'96 Prelude with the manual transmission. The Place Racing adapter bracket must be fastened to the rear crossmember of the '88–'89 Prelude for you to reuse the original rear mount. Three holes must be drilled and tapped for bolting the new bracket into place on the rear crossmember. Fortunately, Place Racing includes the drill and tap set with the purchase of the '88–'89 mount kit. They also include instructions and a template to ensure you position the piece properly. If you have a '90–'91 chassis, the rear mount can remain bolted to the rear crossmember, untouched. Regardless of which year or type of Prelude engine is being swapped, the new rear bracket should be slid into place after the driver-side mount is connected to the engine. You should do all this before you attach the transmission bracket.

Once the rear bracket is in place, tighten all three mounts to the proper specifications. If you use Place Racing's

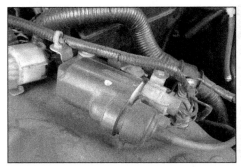

This external distributor coil will come in handy if you're using a USDM H22A1 engine with this type of setup. If so, simply plug the old coil wire into the new distributor.

Unless you're up for quite a bit of welding, these specially designed engine mounts from Place Racing will be mandatory for the H series transplant.

Exhaust System Issues

Your chances of being able to bolt the original exhaust system back up to the new engine's exhaust manifold are pretty slim with engine swaps. Other than a select few engine transplants, odds are that you'll have to do some welding for everything to reattach properly. Problems will arise from donor engines with exhaust manifolds that are significantly shorter or longer than the vehicle's original unit. Some also have mismatched flanges or came without provisions for a catalytic converter. At least one of these issues is going to have to be addressed on most all engine swaps.

A couple of solutions are available for dealing with a large gap between the donor engine's exhaust downpipe and the vehicle's exhaust system. By simply cutting a section of the original exhaust system in half, you may be able to weld a piece of tubing in place to act as an extension of sorts. When you make the cut, be sure to do it on a straight section of the exhaust, forward of any hangers or brackets. Making the cut behind the rubber exhaust hangers will prohibit you from moving the exhaust flange closer to the downpipe of the donor engine.

On the other hand, for donor engines equipped with downpipes that are too long, simply cut a section of tubing out of the original exhaust to move the flange back. Again, try to make the cut in a straight section of tubing that's as close to the front of the vehicle as possible. In all cases, avoid chopping up the downpipe of the engine, as it is usually best to modify the exhaust piping instead.

In certain cases, problems will arise due to an incompatibility between the exhaust flange on the vehicle and that of the new downpipe. In these situations, the proper flange must be welded to the end of the vehicle's exhaust piping. Before laying the weld bead down completely, just tack it initially and reinstall it for a test fit. Once it's fitted properly and welded up, install the proper gasket with the new flange.

Another problem you might find is that some post-swap vehicles will be without a catalytic converter. If you have one of the several Civics that included the converter as part of the exhaust manifold on the old engine, the vehicle will be cat-less unless one is retrofitted into place. Purchase either an OEM unit or a high-flow aftermarket converter, and install it as close to the engine as possible. With the right flanges and gaskets, you can slip the new converter into place by cutting a section out of the original exhaust system. Again, avoid cutting the downpipe, as it is much easier to work with the exhaust tubing.

Remember, you're going to have to allow for engine movement. By welding or bolting the entire exhaust system together without any means of flexibility, you're just asking for a cracked flange or split downpipe. If you use the OEM Prelude flex pipe or the Civic's spring-loaded exhaust bolts, the exhaust system will be able to handle a reasonable amount of engine movement.

Notice how a different flange has been welded onto this pipe, plus it has extended a couple of inches.

These spring-loaded bolts found on many Civics and Integras will keep the exhaust flanges and welds from cracking when the engine moves.

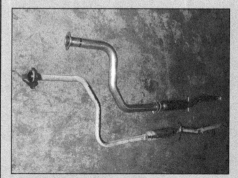

The best place to cut these exhaust pipes would be just in front of the first bend (toward the right in the picture). If you cut too far back, you might mess up the alignment of exhaust hangers.

An alternative to the spring-loaded bolts, flex pipes such as these are often times the only option if a much larger exhaust pipe diameter is being used.

mount kit, the engine will sit in an optimal location, allowing for ample hood clearance and a suitable amount of room between the pavement and the oil pan. Surprisingly, there is quite a bit of clearance in front of the engine as well. With the exception of some minor fabrication involved on the '88–'89 chassis, engine transplants just don't get much more straightforward than this one.

Suspension and Axles Key Points

Since the suspension can just be reinstalled, you have little to do underneath the vehicle except for installing the axles and the shifter mechanism. Due to the uncommon nature of this engine transplant, custom axles are necessary. Measurements can be taken from the seal on the transmission to the innermost portion of the wheel hub. When measured properly, you'll find that the driver

Specially designed axles will be necessary due to the differences in hub diameters. Notice how this axle has the smaller, Integra type of splines.

side measures up just slightly shorter than the passenger side. These measurements apply only when you use the Prelude manual transmission intermediate shaft. Accord intermediate shafts can be used but will require a different left-side axle. Once you have these measurements and know the type of engine you want to install, most competent driveshaft shops should be able to put a pair of axles together for you fairly quickly. Since the older Preludes have smaller hubs, the outer portion of the new axles will look more like Civic or Integra axles, while the inner joints will match the larger, newer Prelude units. For folks who don't trust their measurements or simply

don't want to have axles custom made, Place Racing always has several axles in stock for these swaps.

Shifter Assembly Rules

Although the '88–'91 Prelude uses a cable-style shifter mechanism similar to the late-model Preludes, surprisingly, it isn't compatible with any of the new engines.

If you reattach your old shifter to the new transmission, you may think you've outsmarted the pros, until you realize a slight problem with engaging gears. Due to the difference in cable lengths near the shifter, the arm on top of the gearbox won't be moved all the way into position, resulting in a grinding sound during shifting.

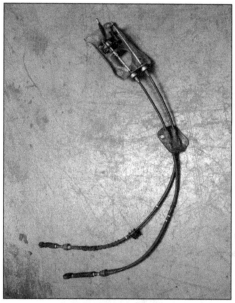

These shifter cables from a late-model Prelude are necessary since the original ones aren't long enough.

This special throttle-cable bracket from Place Racing will allow you to reuse the original Prelude throttle cable. It simply bolts onto the intake manifold.

Avoid this mess by using any complete '90–'97 Accord or '92–'01 Prelude mechanism instead. Do not mix and match cables or shifters, as this will introduce a whole new set of problems. Finish off cables and lines by reattaching the throttle cable using a special bracket from Place Racing. This specially fabricated bracket will allow you to use the original throttle cable, so you won't have to crawl underneath the dash to unclip it.

The original hydraulic clutch line can be hooked back up at this time as well. Since an H-series slave cylinder needs to be installed, you'll need to bleed the hydraulic clutch fluid before you can engage the clutch.

Fuel and Cooling Systems Simplified

Moving on to the fluid system, the fuel-line situation is straightforward. Use a Prelude fuel rail with the inlet port on

In this case, the original Prelude fuel injection feed hose simply reattaches to the new Prelude fuel rail. Be sure to replace the crush washers for the ultimate seal.

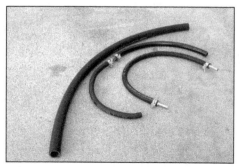

When making a custom fuel line, be sure to use fuel injection hose and hose clamps to ensure a leak-free setup.

In most cases, the original radiator is fine. The H-series radiator hoses connect fine with the original hose clamps because they have the same size inlets and outlets as the radiator.

the passenger side to achieve a factory fit. You might have to get this type of rail from a '92–'96 USDM Prelude if your engine doesn't already have one. Then simply reinstall the original Prelude fuel-injection hose from your vehicle.

Be sure to replace the two aluminum crush washers to prevent any future leaks. The return side of the fuel rail can now be reattached to the original return line.

Since the '88–'91 Prelude radiators feature the larger inlet and outlet water ports, installing the '92–'96 Prelude radiator hoses will be a cinch. When reinstalling the radiator, the original cooling fan can also be retained. There

will be plenty of room to spare between the fan and the new H engine.

A/C and Power Steering Pointers

In addition to cooling the engine down, you can also cool down your interior, if you have the right components. Air conditioning may be retained on the Prelude, but not without a little effort. An H-series A/C bracket is mandatory, but you still can't use the original Prelude air compressor. You'll need a late-model '92–'96 air compressor, and some custom lines.

The same thing goes for those who wish to keep their power steering. You'll need late-model Prelude brackets and pumps for the system to bolt on properly. You can make a custom line by cutting the end off of the original high-pressure hose and brazing or welding on a late-model fitting. Many donor engines that come with power steering pumps still attached will also have this line, since it will be cut at the wrecking yard instead of being unbolted.

If this is the case, all the proper parts for A/C and power steering might be right under your nose. If customizing your own power steering feed line isn't for you, consider taking the proper measurements and having one made.

Sufficient Braking and Suspension

All third-generation Preludes are

standard equipped with front and rear disc brakes, so additional upgrades because of the new engine won't be necessary. The Prelude braking system is surprisingly efficient and needs only some slotted rotors and a set of high-performance pads to improve it.

The same can be said for the BA Prelude's suspension. The difference in weight is comparable to adding a ZC into a Civic. Handling changes will be less than noticeable to say the least.

The H Series is Your Best Bet

The H-series engine swap into the older Prelude chassis is a rather affordable solution for more horsepower, and is the number one route to take as far as many are concerned. Since the B series engine in these cars is inferior and invisible as far as the aftermarket is concerned, turbocharger and supercharger kits aren't always available options. A typical H22A engine costs less than either of these power adders, so you can see why this is the more prominent choice. Providing you use the Place Racing kit and a well-built set of axles, the H-series swap into this chassis will provide long-term, reliable service to this old Honda. With the exception of the bright, powder-coated engine mounts, one would be hard pressed to tell that this vehicle has anything but the original engine in place.

When purchasing a donor H-series engine, be sure to find one with the proper power steering components. You'll need these to make the Prelude line work with the H-series pump.

Once the more powerful H-series engine is installed, a pair of cross-drilled rotors and high-performance pads will be in order. (Photo Courtesy of Progress Suspension)

1990 TO 1997 ACCORD, 1992 TO 1996 PRELUDE, AND 1997 TO 1999 CL

The Honda Accord, Honda Prelude, and Acura CL can best be examined together when referring to engine transplants because of their similar chassis and body characteristics. Thanks again to the engineers and designers at Honda, the similarities between these cars make the art of engine swapping that much easier. Both the Accord and the CL (which are one and the same) share a special relationship with the Prelude, as do the Civic and its big brother the Integra. Like the Civic and the Integra, all late-model Accords and CLs share the same chassis configuration up front with the Prelude. This makes it possible for you to simply bolt in a bigger, more powerful engine. Interestingly enough, although the '90–'93 Accord chassis is completely different externally from the newer '94–'97 Accord, they're similar enough that basic bolt-in engine swaps can be done in much the same way.

Honda Accord Overview

If you look at the Honda Accord from a historical standpoint, there is no doubt that this has been one of the best-selling, most influential cars in the last 20 years. However, looking at it from a high-performance standpoint, you'll hardly get the same impression. Even though the Accord is worthy enough to be among the vehicles in this book, it just hasn't taken off like the Civic or Integra.

There are two underlying reasons for the Accord's lack of success in the aftermarket. First, the Accord has always been among the most expensive that Honda had to offer, making it unattainable for the typical younger import enthusiast. Second, the Accord just isn't designed for speed or racing.

With this in mind, when most people think of the Honda Accord, probably the last idea to come to mind is a high-performance engine swap. With plush interiors, the best amenities Honda has to offer, and some models approaching a curb weight 1,000 pounds more than a CRX, the Accord does seem a little unlikely to be the recipient of a 200-horsepower Prelude VTEC engine. Even if it isn't the most popular vehicle

From this angle, you would be hard pressed to tell that an engine swap has even been done. That's one of the nice things about the Accord chassis and Prelude engines.

to receive an engine swap, you be surprised what can be accomplished with the '90–'93 and the '94–'97 chassis Accord.

'90–'93 Accord Offerings

With the introduction of the '90–'93 CB Accord chassis, only three available trim levels were offered, but a Special Edition (SE) was sold only in '91 and '93. The most expensive of the bunch is the EX model, followed by the more stripped-down LX and DX. The pre-VTEC F22A1 engine could be found in all '91 Accords. With the introduction of the '92, the F22A4 and F22A6 became available in the EX and SE. Despite their different engine codes, these three engines are all very similar. Available body types for the CB-chassis Accord include the two-door, four-door, and wagon models.

This model Accord was introduced in 1990, but it really didn't receive much attention as far as engine swaps are concerned until the late 1990s.

The featured SOHC F-series engines aren't the best engines to modify, especially considering how easy it is to install a DOHC H series.

'94–'97 Accord Offerings

Upon the unveiling of the fifth-generation '94–'97 CD-chassis Accords came the first SOHC VTEC F-series engine. Available only in the EX and SE trim levels, the F22B1 put out 145 horsepower and was the most powerful Honda SOHC four-cylinder engine of its time. Lowlier LX and DX models received the non-VTEC version known as the F22B2. Both of these engines could be found in the coupe, sedan, or wagon versions of the fifth-generation Accord. Of all the Accords listed, four-wheel disc brakes could only be had on the top-of-the-line EX and SE models starting in 1992. Since most Accord components are interchangeable with one another, rear disc brake swaps are a cinch on any '90–'97 model.

With the introduction of the '94 Accord, folks started to take notice of the H series swap. This chassis is preferred among all Accords for the H swap.

Honda Prelude Overview

Since some Preludes came factory-equipped with H22A1 engines, it should be no surprise that any of the H-series engines can be retrofitted into the lowlier Prelude models.

Introduced in 1992, the BB chassis Prelude has always been a benchmark of what Honda high-performance stood for. However, as one of the heaviest and most-expensive four-cylinder Honda/Acuras, fourth-generation Preludes with engine swaps are a rare sight. The price tag is much higher than most Honda enthusiasts are willing to spend, so there aren't too many heavily modified Preludes. The fact that they weigh in at around 2,900 pounds and came with a completely useless back seat doesn't help much either.

'92–'96 Prelude Offerings

As with the '90–'93 Accord, the same F22A1 engine may also be found underneath the hood of all '92–'96 Prelude S models. Being available only in the two-door coupe form, the Prelude was only offered in three different trim levels; the upper two models were the Si and Si-VTEC versions. Rare special edition and 4WS models were also available for a short time featuring the H23A1 engine, which is found in the Si version. The top-of-the-line Si-VTEC model was equipped with the ever-so-opular H22A1.

Acura CL Overview

Introduced in 1997, the Acura CL is more or less an expensive, more-refined version of the Accord. The CL offers the same engine, chassis, and engine bay as the '94–'97 Accords. If you have the optional six-cylinder engine available on top-of-the-line models, it wouldn't be

These '93–'96 Preludes are where the H series is found. These are also popular swap candidates, as several models were equipped with lesser SOHC and non-VTEC engines. (Photo Courtesy of Progress Suspension)

very sensible to go swapping this drivetrain out for that of a four-banger Prelude. Since they share similar engine bays, it goes without saying that swaps for the Accord and CL can be performed in the same manner.

Though it certainly doesn't happen often, the CL can be fitted with an H series just as easy as any of the Accords.

'97–'99 CL Offerings

As with the Accord, the Acura CL base model was equipped with the SOHC VTEC F22B1 engine. A new and more powerful SOHC VTEC F23A1 engine found its way under the hood of the '98–'99 CLs. Regardless of whether the CL is equipped with either of the two four-cylinders or the more powerful six-cylinder, the CL is only available in coupe form, with base or premium trim levels. You can expect to find four-wheel disc brakes standard an all CLs available.

OBD I Vehicles

These vehicles' electrical systems overlap OBD I and II. The '90–'95 models are OBD I. These vehicles should remain that way unless an OBD II computer is absolutely necessary. This situation is interesting if you think about it. The '90–'91 Accords are one of the only OBD I Honda vehicles manufactured before 1992.

When using an OBD I ECU on '90–'95 vehicles, simply plug the ECU into place and you're ready to go. For OBD II conversions, you can find the proper OBD II connecters in any '96–'01 Accord or Prelude, or '96–'98 Civics and Integras. Be careful not to use plugs from the '99–'01 Civic or Integra; they're quite different from the earlier pieces. Whichever plugs you get, they'll

need to be grafted onto the '90–'95 underdash harness one by one. You'll need to consult the service manuals for your vehicle and your ECU of choice to make the proper connections.

Once you've soldered the connectors into place, your ECU of choice can be plugged in. If you buy an OBD II engine, you can still use an OBD I ECU, providing you make the proper modifications to the engine wiring harness. If you want an OBD II ECU, remember to only use the JDM '96–'01 versions. These ECUs may be used if you add the proper connectors to the underdash harness of the vehicle. Again, with several OBD I ECUs available, this is usually unnecessary and will only cost you more work.

OBD II Vehicles

The second situation involves the OBD II '96–'99 models. In this situation, you can either retain OBD II or convert to OBD I. Make this conversion with the help of an adapter harness from Place Racing or Hasport Performance. When using OBD II engines on these vehicles, most of the OBD II ECUs listed above may be installed with the exception of the '97–'01 USDM units. OBD II engines can be used with the OBD I ECUs as well, providing that the vehicle wiring harness is modified for the new connectors. Rather than shopping for electrical connectors at the local wrecking yard, plug-in adapter harnesses are available.

H-Series Engine Swap

With so many similarities between the Accord, Prelude, and CL, it's difficult to discuss engine swaps for one and not the others. All three vehicles follow many of the same guidelines and principles concerning engine transplants. With the introduction of the 1994 lineup, Honda introduced an Accord that was alike in many ways to the sportier Prelude, introduced in 1992. Thanks, again to the folks at Honda; the Prelude and Accord chassis are very similar, allowing for quick, easy, inexpensive engine swaps. The same situation occurred when Acura unveiled the CL in 1997. Once the engine swap craze began, it was clear that the older fourth-genera-

tion Accords were equally compatible.

Unlike the relationship between the Civic and Integra, most of the similarities lay in the chassis. The Accord and Prelude share many engine and drivetrain parts as well. When comparing the later models just mentioned, you'll find almost identical transmission cases, engine mounts, engine brackets, engine blocks, oil pans, and cooling hoses and other parts.

Any time you have a situation like this, it will not only simplify the swap, but also keep the cost down at the same time, since many of the parts you need are already right in front of you. Someone who knows their way around one of these vehicles is probably more than qualified to perform this transplant. The swap process should consume no more time than a standard engine replacement, so a typical H-series transplant should take no more than a long day or two in the garage.

H-Series Engines and Transmissions

To select a donor engine for any of these vehicles, you first need to know the OBD status of the car. Although not as important as with some of the other swaps in this book, selecting an engine with the same OBD II level as the vehicle will usually save you both money and time. The '93–'96 USDM Prelude Si-VTEC is certainly on the list of the most popular of engine swaps for these particular vehicles. These 190-horsepower H22A1s are relatively easy to find in local junkyards and make a perfect addition to any one of these chassis. The similar JDM H22A produces slightly more power and can be found from the same type of vehicle in their JDM versions. Of course, the '96 OBD II versions of these engines are best suited for the '96 and newer OBD II vehicles.

Another possible Prelude VTEC engine swap candidate is the '94–'97 JDM Accord SiR. The fact that this engine was originally offered in an Accord chassis demonstrates just how compatible these cars are. The 220-horsepower JDM H22A in the '97–'01 Prelude S Spec is most coveted donor engine of all. EDM H-series engines are

Late Style Auto to Manual Conversions

If your engine-swap vehicle was unlucky enough to start life as an automatic, converting it to a manual transmission won't prove too difficult. Fortunately for Hondas and Acuras that could have been equipped with the optional hydraulic-style transmission, components are available to make these conversions happen. They are available from both the aftermarket and the OEM. Before you set your donor engine in place, you must first add hydraulic-operated clutch components to the chassis. The studs and bolt holes are already there, so simply bolt a clutch master cylinder, a clutch fluid reservoir, and all of the necessary hydraulic clutch lines into place.

However, in most cases, you'll need to drill some holes; there are usually clear indications on the firewall as to where they need to be. You'll also need to bolt up a pedal assembly to finish things off. All of these parts can be found from any USDM manual-transmission version of your vehicle of choice. Of course, you'll need all of the manual transmission mounts, axles, and a shifter assembly of some sort as well. Before installing the shifter, screw an aluminum plate down over the automatic

You would never know that this pedal assembly wasn't always a part of this formerly automatic Civic. After removing the old unit, this new manual transmission version bolted right up.

This old automatic shifter assembly can be tossed.

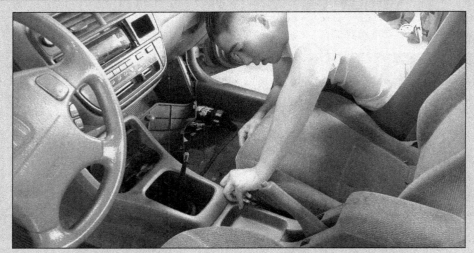

Once the interior is reinstalled, nobody will ever know it was once an auto.

shifter hole. Cut a hole in this plate just small enough to fit the new shifter through. This will help eliminate any fumes from entering the vehicle through the auto shifter's larger hole.

When dealing with the '92–'95 Civic and '94–'01 Integra chassis, you'll need a special mount from Place Racing or HCP Engineering to use the B-series manual transmission. Due to differences in the passenger-side pocket mounts, you can install this special bracket instead of cutting the auto version off and welding the manual one into place.

Keep in mind though, if you're converting to an H series, cutting and welding will be mandatory because a mount is not available. Accords from '90–'93 pose a similar problem for their owners, but you can simply space the transmission down a bit with several washers. This may be done in lieu of welding on the manual-style frame-rail mount.

Installing a K-series manual transmission into the '01–'03 Civic chassis, or converting any '96–'00 Civic is simply a bolt-in process. Converting the '94–'97 Accord or '97–'99 CL is just as easy. If you find yourself installing the clutch master cylinder on these last two chassis however, be careful because the firewall is especially weak on the automatic vehicles. You'll need to add reinforcements under the dash; otherwise, the master cylinder will deflect several millimeters when you press in the clutch pedal.

To finish up your conversion, you'll also need to do a bit of wiring. First off, in order for the reverse lights to work later on, the wires will have to be removed from their location near the shifter assembly and rerouted to the reverse-light sensor on the

gearbox. You also have to fool the car into thinking that it is constantly in park. Do this by removing the shift lock solenoid from the ignition, so you can remove the key from the vehicle once it's turned off.

If you want to keep the automatic in the Civic or Integra, you'll find that there are a couple different options. For any B-series swap into a Civic or Integra, simply reuse all the original mounts. The same thing applies to Accords, CLs, and Preludes going H series. Although most would highly recommend the rare JDM automatic GSR transmission for the B series, the original RS and LS gearbox from the Integras can be reused in lieu of the high-performance version. H-series swappers hoping for an automatic transmission conversion might want to consider the automatic Prelude VTEC transmission from Japan. In most cases, whenever you use an automatic gearbox, be sure to use the corresponding ECU for the transmission. This will provide you with the proper shift points for optimum performance.

This HCP Engineering transmission mount will save you from cutting and welding.

This JDM H22A engine is a perfect donor engine for these vehicles, as it is a direct bolt-in fit.

available as well but will be much tougher to come by here in the United States. Such engines include the H22A2 found in the '94–'97 Prelude 2.2i and the H22A7 of the '94–'97 Accord Type R.

Although the '97–'01 USDM Preludes do offer a 200-horsepower H22A4 engine, it is much less compatible than the other available engines, and so should be avoided. These engines are basically the same as other H22As in dimensions and outward physical characteristics, but they're also equipped with the ATTS transmission and an incompatible ECU. Using the ATTS transmission involves a lengthy wiring procedure and custom mount fabrication, which are simply unnecessary. With the vast array of compatible Prelude and Accord transmissions on the market, the ATTS unit is something that you don't need to mess with. In addition to the incompatibility of the gearbox, the '97–'01 USDM Prelude ECUs are equipped with an internal anti-theft device that renders them useless outside of their original vehicle. If you're using a '97 or newer H engine, just remember that a lot of the parts that you'll be getting with your purchase will be useless.

Moving on to other compatible engines, a few choices are available for those not interested in Honda's variable camshaft timing. Most popular of all H23As, the H23A1 may be found in the '92–'96 USDM Prelude Si. The same engine may also be found in the JDM version of this vehicle and is referred to as simply the H23A. The EDM has its own version of this engine known as the H23A2, found underneath the hood of the '92–'96 Accord 2.3i. Although not exactly a part of the H series family, the DOHC F22B is a cheaper, more popular alternative to the H23A, found in the '92–'96 JDM Prelude Si.

Use any 1996 or older transmission for any of these engines. In fact, any '90–'97 Accord or '92–'96 Prelude transmission will work just as well. The Prelude VTEC transmission is probably the most popular from a performance standpoint, with its close-ratio gears and optional limited-slip differential in the JDM version. For those more concerned with gas mileage or just a lower initial cost, any of the Accord or non-VTEC Prelude transmissions will work. You can even reuse your old CL transmission for that matter.

ECUs and Wiring for the H Series

In most cases dealing with the H series, it's usually recommended to use an OBD I ECU for performance and availability reasons. However, due to emissions regulations in some states,

you might have to go with an OBD II engine and ECU. That will leave you with a few possible scenarios when dealing with the vehicle's electronics.

The first scenario involves using a '90–'95 vehicle and swapping in a compatible OBD I engine. This route will require the least wiring providing you're using the proper OBD I computer. However, this part is easy. For the VTEC engines, use the P13 ECU; for the non-VTEC engines, you'll want the P14. Both of these ECUs can be found in the '92–'95 USDM or JDM Preludes.

A second scenario involves a '90–'95 vehicle and an OBD II engine. You have two routes you can take to use the newer engine in this vehicle. First, using the OBD II ECU necessitates an ECU adapter harness to compensate for the differences in electrical connectors. Assuming you're using a JDM (not USDM) ECU unit, then this will certainly work. Although the '96 USDM versions don't have the built-in immobilizer system, they are equipped with USDM-only emissions components that you'll want to avoid. You'll have to find plugs from another vehicle that match your OBD II ECU and solder them into place underneath the dash of the project car either way. A second solution to this problem will require you to use an OBD I ECU on the OBD II engine. This works just as well and is actually much easier.

The final scenario that you may encounter involves an OBD II engine in an OBD II vehicle ('96 Prelude, '96–'97 Accord, and any CL). You can keep your OBD II electronics system and

You need to make several modifications to the engine wiring harness. Once the harness is removed from the original engine, you can lay it out on your workbench and take care of business.

simply plug in the new ECU. Again, remember to avoid the USDM ECUs that are equipped with the immobilizer.

Regardless of the type of engine you select, you'll need to make several wiring modifications to the original engine's wiring harness. Position the harness onto the new engine starting with the injector plugs. Continue attaching all the electrical connectors, lengthening and shortening the wires as necessary. You'll probably have to lengthen the wires for the intake air temperature sensor, the map sensor, the vehicle speed sensor, and the oxygen sensor. You may also find

you need to lengthen the wires for the fan switch, idle air control valve, and engine coolant temperature sensor.

In situations where you'll be crossing over to an engine of a different OBD type, the distributor plugs will also need to be swapped out. With most DOHC Prelude engines, you'll also

have to add wiring for the intake air bypass valve and the knock sensor. To ensure a factory-type fit, the alternator wires on the original harness must also be lengthened a few inches or replaced with the Prelude section.

Continue by lengthening the oil-pressure sending unit wires and replacing

This external coil on the '94–'97 Accord is retained so that it may be reused with the external-coil distributor of the USDM Prelude H22A engine.

Cut the alternator wiring off of the donor H-series engine harness for a factory-looking fit, as far as the wiring harness is concerned.

Intake Air Bypass Valves

Intake air bypass (IAB) valve, otherwise known as secondary intake butterflies, can be found on all '94–'01 Integra GSR engines and all '88–'01 Prelude DOHC engines. Equipped with two separate intake paths inside the manifold, the dual-stage intake allows the incoming air to follow the path that is most favorable depending on the engine's RPM. Optimum power and torque is realized by opening and closing the IAB valve. By closing the valve, additional torque can be found in the lower RPM range. Since it isn't directly related to the emissions system, often the IAB is left unwired without any ill effects. In this case, the valve will be in the open position and won't pose any real restriction. Many folks choose to use computers that aren't even equipped to recognize the IAB system. This is perfectly acceptable as it does not affect drivability in any way.

This late-model Integra GSR intake manifold is equipped with the secondary intake system. The actual IAB is the disc shaped unit shown in the lower right.

Notice the stacked manifold design. These secondary-type intake manifolds are actually three pieces bolted together.

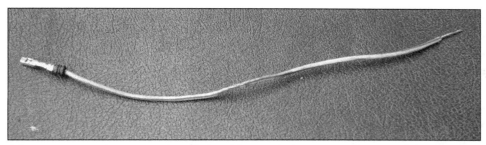

When you move plugs around and add wiring to the harness, using pins like these in the plugs will give you a more factory look as opposed to cutting and soldering.

the plug from the donor engine's harness. The same process must be performed on the EGR valve if a Prelude computer is to be used.

For VTEC-equipped engines, you must add wiring for the VTEC pressure switch and the VTEC solenoid. You'll need to run these wires down to their appropriate pin locations on the ECU to ensure that the engine functions properly under VTEC operation. Vehicles already equipped with VTEC need only have their wires lengthened to reach. Select JDM OBD II engines aren't equipped with a VTEC pressure switch, so you don't need to wire one into place. Keep in mind, however, that if the ECU you're using is looking for a VTEC pressure switch, then you must install one, even if it isn't on the engine. You can find the proper switch from another engine.

A few final wiring modifications are necessary only for the '94–'97 Accord. You need to switch pins A6 and A11 on the ECU. This will help ensure that both your oxygen sensor and EGR valve work properly.

Installing the H Series

Installing an H-series DOHC engine into an Accord, Prelude, or CL chassis is fairly simple. Since you can use a hodgepodge of OEM Honda mounts and components, people unaware of the differences between Honda engines will never even know the swap has taken place.

Begin by placing the driver-side engine mount in the frame bracket. You can reuse the original rubber mount for all Preludes and '90–'93 Accords. You'll need a mount from a '92–'96 Prelude for the newer Accord and CL chassis. You'll notice that the mount doesn't quite fit

into the pocket on the newer Accord and CL. Shave a small amount of metal off the bushing inside the mount, and slip it into place.

The rear engine and transmission brackets may all be reused except if you have an earlier Accord (instead, use the stock one). Of course, this is assuming that they're manual transmission units

Install this Prelude driver-side mount on the left-side pocket mount in the '94–'97 Accord. Before you can do this, you'll need to cut down the bushing protruding out the right side.

Reuse this Accord front mount along with its corresponding bracket. The stiffer Prelude units are not compatible with the Accord crossmember.

and that you use the original rubber mount that they were attached to. This means that if you're going to use a Prelude rear engine bracket on your Accord, then you must also replace the rear rubber mount with a Prelude unit. Sometimes this is the desirable way to go on the Accords and CLs, as the Prelude components are stiffer. Of course, this isn't mandatory, and you can just reuse your original components.

Moving on to the front mount, you must reuse the Accord and CL units on these chassis, as the stiffer Prelude piece won't bolt up. As you'd expect, the Prelude front mount must be reused on the Prelude.

Last, reuse your right-hand transmission bracket with your new transmission, assuming the car is already a five-speed.

With the new H engine in place, you'll soon notice just how well it sits under the hood. What else would you expect, knowing that H engines are already available in the higher-end USDM Preludes and the top-of-the-line overseas Accords? Because Honda designed these chassis for the more stout H engines, the swap will prove to be a straight bolt-in affair without the need of any cutting or persuading of the frame. Most H-series engines have the same mounting characteristics and oil pan of the previous SOHC engines, so there will be zero issues with ground clearance.

Notice the tight fit between the radiator and the alternator. If you push it a bit farther backward, there'll be just enough room between the two.

Axles, Shifters, Cables, and More

With the engine bolted into place, finishing underneath the car will take no time at all, thanks to the compatibility of

Simply leave your cruise control in the factory location, as it will not need to be removed.

Rather than mounting it on the firewall, the owner chose to mount this MAP sensor to the valve cover for a tidy look.

Prelude and Accord parts. Reattach the original suspension and some '92–'96 Prelude axles. Providing you use an intermediate shaft from a '92–'96 Prelude five-speed, the manual transmission axles will slide right into place without any modifications.

In all cases, the original shifter mechanism can be reused on any of the aforementioned gearboxes. Since you didn't have to remove the shifter from the car to remove the engine, you should only need to reattach the cable to the top of the transmission with new cotter pins.

Moving on to other cables, replace the throttle wire with one from any '93–'96 Prelude VTEC. It should work with the proper bracket, which should already be on the donor engine. If you want cruise control, the original smaller Accord or CL throttle body must be reused, along with its cable. You'll need to fabricate a special bracket to attach the cable to the new intake manifold of the Prelude engine.

If you aren't using an Accord throttle body, an external MAP sensor must be installed on the firewall and plugged into the harness. This is because the MAP sensor is located on the firewall of Prelude VTEC models, as opposed to the top of the throttle body on the Accords and CLs.

Run a piece of vacuum line to the sensor for it to work properly. External MAP sensors can be found on the '88–'91 Civic or '90–'93 Integra, for example.

Fluid Systems How-To

Depending on your vehicle of choice, you have a couple of options when dealing with the fuel system. The fact that the newer Accord and CL fuel filter is located

Fuel Systems

With so many different fuel lines, filters, regulators, and fittings found on the vehicles in this book, it's a common mistake to get them mixed up. Most Honda and Acura vehicles feature a return-style fuel system that recycles unburnt gasoline back into the gas tank for future use. This system usually consists of a gas tank, a fuel pump, fuel-injection line, a fuel filter, more fuel-injection line, a fuel rail, a fuel regulator, more fuel-injection line, and eventually a final return back into the gas tank. Of course, the fuel injectors lie just past the fuel rail, providing the necessary fuel to the engine.

There is much confusion about the fuel injection line that connects the fuel filter to the fuel-rail feed port. This is usually connected via a piece of 8-mm fuel injection hose with a banjo-style hose end crimped to each end, direct from Honda. One will find that the fuel filter has a female hole threaded into it for the insertion of the proper-sized bolt, which fastens one end of the banjo to the filter. The other end of the banjo slides over a hollow, threaded stud (some use a bolt instead of a stud) on the end of the fuel rail accompanied by one of two special closed-end sealing nuts. This is where much of the confusion lies. The proper banjo hose end must be used. Which one depends on the fuel rail and the sealing nut that goes with it. Mismatching the hose ends and sealing nuts will result in a dangerously lean condition, if the car manages to start at all.

As with most problems in this book, there is usually more than one solution. The easiest way to solve this particular problem is to visit the local Honda or Acura dealership and simply order the correct hose specified for each swap. It's important to order the hose that corresponds with the engine and not the vehicle. Where the hose attaches to the fuel rail is most critical, as the connection to the fuel filter is similar on both styles of hose ends. The other solution involves simple cutting and splicing techniques and will cost significantly less than purchasing a new hose assembly. In order for this procedure to work, there must be a portion of the injection hose still attached

This damper pulsation nut assembly is more common to newer vehicles, although it can be found on select Si models all the way back to 1986.

This standard cap nut assembly is found on many base-model Hondas and most older vehicles.

to the fuel rail on the donor engine. With the injection hose on the car cut in half, and the injection hose on the donor engine's rail cut, simply add a new piece of fuel-injection hose in between the two with the proper splicers. The most effective and most permanent splicers available are the 5/16-inch brass barb connectors that can be found in the plumbing section of most hardware stores. The procedure can be finished off with fuel-injection hose clamps on each connection. Although buying the new hoses is the preferred and the most professional, often these special OEM hoses are very expensive and may not be an option for some.

on the driver side of the vehicle poses a small problem when hooking things back up. The easiest solution consists of installing a fuel rail from a '97–'01 Prelude. Since the inlet on most Prelude fuel rails is on the passenger side of the vehicle, you'll have to lengthen the fuel injection feed hose to reach certain donor engines' fuel rails. Installing the newer fuel rail with the inlet on the driver side will provide you with a leak-free, factory-looking fit. Nevertheless, before purchasing the '97–'01 fuel rail, wait and see what comes with the engine you purchase, because certain JDM engines built before 1997 do include this newer-style rail.

In contrast, to finish off the fuel system on Preludes and older Accords, simply reattach the original fuel injection feed line to the H-series fuel rail for a factory-type fit. It's easier on these models since the fuel filter is located on the passenger side.

The cooling system can be reinstalled with a minimal amount of effort. In all cases, the original radiator should be reused along with its cooling fan. Upper and lower radiator hoses found on any of the '92–'96 Preludes will suffice for all H-series transplants into any one of these chassis.

If you're swapping into either a '92–'96 Prelude or '90–'93 Accord, you can reuse the radiator hoses if they're in good condition. Since all of these vehicles' radiators feature the larger inlets and outlets, you can reuse the original hose clamps in all cases as well. Use the heater hose from your H-series engine to connect to the cylinder head, as they're just a bit longer.

A/C and Power Steering Components

With all of these Accords, CLs, and Preludes equipped with air conditioning and power steering right from the factory, it's understandable that you might want to retain these features. If you do, using the proper components will make the installation a breeze. Use the original Accord or CL A/C bracket and compressor on Accords and CLs. You can also reuse all the original lines. The same thing goes for any DOHC H-series swaps into the Prelude chassis, just reuse the Prelude components, and you're good to go.

Retaining your power steering in these vehicles isn't much more difficult. Use the H engine's power steering pump bracket; it will bolt right on to the engine block with plenty of room to spare. Due to differences in the power steering pres-

Notice the air conditioning lines running alongside the front of the radiator. There is also plenty of clearance in regards to the cooling fan and the condenser fan as well.

You can install a Prelude upper radiator hose to work with the original Accord radiator.

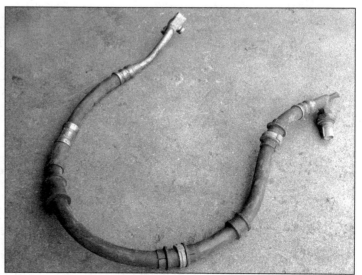

Often, power-steering hoses such as these may be acquired from a junkyard engine.

sure line, you'll need to cut off and replace the fitting on the end near the pump with that of the H22A or H23A.

After you've brazed or TIG welded the new end into place, bolt the new line onto the pump for a factory finish. Don't skimp and try to make this connection with hose clamps. Try it and you'll find a sizeable amount of power steering fluid covering your engine bay in a hurry.

Further Improvements

With a minimal amount of additional weight underneath the hood, negative effects of understeer will be hardly noticeable. If you do notice some understeer, you can counteract it by upgrading your suspension. Progress Suspension makes some performance-minded lowering springs that, along with some upgraded front shocks, will alleviate some of these problems.

If you want to hop up your brakes for the utmost safety, just add a pair of larger slotted rotors and a set of high-performance pads. The rear drum brakes found on some Accords can be converted to discs using parts from the '92–'97 EX and SE models. These will bolt right in and give you the additional stopping power that only four-wheel disc brakes can provide.

The Best of Both Worlds

For those who wish for the best in performance and luxury, sometimes a Civic or an Integra just isn't going to cut it. Although none of the vehicles in this section can compare to the all out acceleration of a feather-light CRX with an H series, for some folks there is more to life than speed. Such things include the reliability and peace of mind that you'll have when you've completed a swap using nothing but genuine Honda components. No customizing, no one-off parts to worry about, just pure Honda reliability. It is a lot different from a stripped-down CRX missing its carpeting and rear interior.

An OEM fit, smog legal, and plenty of power. For folks who don't know exactly what they're looking at, you would be hard pressed to think this H22A was anything other than stock.

Power Steering and Components

If removing the power steering in hopes of a lighter chassis isn't for you, in most cases the original system can be easily retained. In most cases, retaining power steering becomes a bolt-on affair by swapping over the corresponding pump, belt, and brackets from the donor engine. In the most complex scenarios, you'll either have to make a custom feed line or modify the original one by means of welding or brazing. In order for the vehicle's line to be compatible with the donor engine's pump, you may have to swap the hose end out. For vehicles without power steering, a power steering rack from a corresponding vehicle can be swapped into place with all the proper components. Folks who'd like to eliminate the power-steering system the proper way can do the opposite.

Ideally, the right way to get rid of your power steering is just to swap in a manual steering rack, like this one.

In this situation, a combination of the correct Integra components was installed for a factory fit.

MALFUNCTION INDICATOR CODES

Imagine how easy life would be if all problems could be identified immediately and solutions were listed in a handy manual. Unfortunately, life doesn't work that way, but when it comes to your Honda or Acura, you might be so lucky. Through the interpretation of the malfunction indicator lamp (otherwise known as check engine light) and following the troubleshooting process detailed in the Honda/Acura service manuals, the solutions to many engine problems can be identified faster than ever before.

Before ECU trouble codes can be read however, you will have to jump the computer on some vehicles first. On all '92 and up models (and '90–'91 Accords), the connector that must be jumped must be found underneath the dash. The connector is located near the ECU on Civics and Integras and behind the center console on Preludes and Accords. Its brown and black wires can identify it and it looks like it isn't plugged in properly.

You need to short this plug out with a piece of wire or a paperclip. Simply fold the "tool" into a U shape and insert it into the plug. When you do this properly and the ignition is in the on position, the check engine light will light up in the gauge cluster.

You can skip this process for most Hondas and Acuras built before 1992. Trouble codes for these vehicles can be read by looking through the small window on the center of the computer.

Don't bother trying to read trouble codes at the gauge cluster as this light won't send any type of signal there unless wired up to do so.

Regardless of which process you follow, interpreting trouble codes is like reading Morse code. Both types of computers will display two different types of blinking lights. Codes 1 through 9 will be indicated by short individual blinks. Codes 10 on up will be indicated by a series of long and short blinks. The number of long blinks represents the first digit and the number of short blinks represents the second. Several codes may be present at once, so be sure to view the cycle at least twice to make sure you don't miss any. A list of codes appears on the top of the next page.

The connector is located just behind the glove box on the '94–'01 Integra. Use a paper clip or piece of wire to jump the two terminals in this connector if you want to read the trouble codes on the dash.

Honda and Acura Trouble Codes

0	engine control module	31	automatic transmission B signal
1	heated oxygen sensor #1	41	heated oxygen sensor heater
2	heated oxygen sensor #2	43	fuel supply system
3	manifold absolute pressure (MAP) sensor	45	fuel system too rich or too lean
4	crankshaft position sensor	48	lean air/fuel sensor
5	manifold absolute pressure (MAP) sensor	54	crankshaft speed fluctuation sensor
6	engine coolant temperature sensor	58	top dead center position sensor #2
7	throttle position sensor (TPS)	61	primary oxygen sensor
8	top dead center position sensor	63	secondary oxygen sensor
9	number-one cylinder position sensor	65	secondary oxygen sensor heater
10	intake air temperature sensor	67	catalyst system efficiency
12	exhaust gas recirculation (EGR) valve	70	automatic transaxle
13	barometric pressure sensor	71	cylinder-one or random misfire
14	idle air control (IAC) valve	72	cylinder-two or random misfire
15	ignition output signal	73	cylinder-three or random misfire
16	fuel injector	74	cylinder-four or random misfire
17	vehicle speed sensor	75	cylinder-five or random misfire
19	automatic transmission lockup control valve	76	cylinder-six or random misfire
20	electrical load detector	80	exhaust gas recirculation valve (EGR)
21	VTEC electronic control solenoid valve	86	engine coolant temperature sensor
22	VTEC electronic pressure switch	90	evaporative emission control system
23	knock sensor	91	fuel tank pressure sensor
30	automatic transmission A signal	92	evaporative emission control system

Smog Certified Engine Transplants

Examples of previously certified engine transplants are listed below. This isn't a complete listing of what is possible, but it does cover those that have been done. Notice that you don't see the ZC or Civic/H-series engine transplants, as these are not smog legal.

1984–1987 Civic/CRX	D series (ZC style)	**1992–1993 Integra**	B series (OBD I)
	B series (OBD conversion may be required for certification)	**1994–1995 Integra**	B series (OBD I)
		1996–2001 Integra	B series (OBD II)
1988–1991 Civic/CRX	D series (SOHC) (OBD conversion)	**1986–1989 Accord**	B series (OBD conversion may be required for certification)
	B series (OBD conversion may be required for certification)	**1990–1993 Accord**	H series (OBD I)
1992–1995 Civic	B series (OBD I)	**1994–1995 Accord**	H series (OBD I)
1993–1995 Del Sol	B series (OBD I)	**1996–1997 Accord**	H series (OBD II)
1996–1997 Del Sol	B series (OBD II)	**1988–1991 Prelude**	H series (OBD conversion required)
1996–2000 Civic	B series (OBD II)	**1997–1999 CL**	H series (OBD II)
2001–2003 Civic	K series (OBD II)	**1992–1995 Prelude**	H series (OBD I)
2002–2003 Civic Si	K series (OBD II)	**1996 Prelude**	H series (OBD II)
1986–1989 Integra	B series (OBD conversion may be required for certification)		
1990–1991 Integra	B series (OBD conversion may be required for certification)		

SMOG CONCERNS

Emissions, hydrocarbons, smog tests; nobody wants to talk about it, but if you're planning an engine swap, then it has to be addressed. It might be the last thing high-performance enthusiasts want to talk about, but avoiding the topic may get you into trouble down the road. Many states have different laws concerning what you can and can't do as far as engine transplants are concerned. Before jumping into any type of engine conversion on your vehicle, it's always best to first check with your local DMV to find out what the law says. Unfortunately, for folks who live in stricter states such as California, your engine swap choices are very limited (see chart on page 122). Some basic guidelines regarding the certification process of engine transplant vehicles that apply to most states are listed below:

1. Donor engines and emissions equipment must be of the same year of the vehicle or newer.

2. Be sure to have a purchase receipt for the engine and transmission before going through the certification process.

3. USDM engines are always easier to have certified compared to their JDM counterparts.

4. Vehicles originally bought and registered in California may only swap in a California registered engine. This applies to some other states as well. However, JDM or any other engines may be used as long as the engine block code is the same as the engine block code on the USDM engine. This is why JDM B16A engines are legal transplants; they refer to the USDM Del Sol DOHC VTEC B16A3. However, if a B16A is to be certified into an OBD 0 chassis, plan on a full OBD I conversion, since the B16A3 is OBD I. OBD 0 B16As do not exist as far as the state referee is concerned. Many times, the vehicle will be required to install the Del Sol intake manifold and exhaust manifold in addition to the entire OBD conversion.

5. Emissions equipment from the donor engine must be installed on the vehicle along with the new engine. This includes the catalytic converter, oxygen sensors, and evaporative emissions components.

6. Vehicles receiving an engine that is of a newer OBD type must upgrade to the OBD level of the new engine. This includes the computer and all of the electrical components. Although the backdating of electronics and downward OBD conversions are mentioned throughout this book, they are illegal in several states.

7. Once you visit a state referee and pass a visual and emissions test, you'll be given a certification sticker. From that point on, you may have your vehicle smog tested in a normal manner at any smog station.

This '89 CRX DX, with its '90 Integra B18A engine under the hood, is emissions legal in even the strictest states. You might be surprised at what you could do, even if you play by the rules.

The exhaust is one area you need to pay attention to if you want to stay emissions legal. Don't "forget" the converter.

FINDING
MAINTENANCE PARTS

Once a JDM donor engine is selected and installed, it can be difficult to find maintenance and replacement parts. More often than not, these parts are waiting for you at your local Honda or Acura dealership. With most all the parts you need right under your nose, sometimes half the battle is simply knowing what to ask for. Use the handy nearby chart next time you find yourself needing OEM parts for your donor engine.

Listed in the nearby table are common JDM engines for which some folks aren't exactly sure where to find common replacement parts. When viewing the chart, keep in mind that this is for common maintenance parts and not for engine overhaul kits and such. More complex and internal items may be found on several different vehicles or may not be found on any USDM vehicle at all.

The basic electrical parts referred to here may include spark plug wires, distributor coils, and various sensors.

Mechanical parts referred to in this chart include water pumps, timing belts, and gaskets, just to name a few.

Keep in mind that although in some cases several USDM vehicles will provide the necessary parts for these JDM engines, the ones listed below are the most popular.

Year	Engine	Part Type	Best USDM vehicle choice for parts
1988–1991	B16A	electrical	1990–1991 Integra
1992–1995	B16A	electrical	1994–1995 Del Sol DOHC VTEC
1996–2000	B16A	electrical	1996–1997 Del Sol DOHC VTEC
1998–2000	B16B	electrical	1999–2000 Civic Si
1996–2000	B20B	electrical	1996–2000 CRV
1985–1987	ZC	electrical	1986–1989 Integra
1992–1995	B18B	electrical	1992–1995 Integra RS, LS, and GS
1996–2001	B18B	electrical	1996–2001 Integra RS and LS
1996–2001	B18C	electrical	1996–2001 Integra GSR
1994–1995	B18C	electrical	1994–1995 Integra GSR
1996–2001	B18C/R	electrical	1997–2001 Integra Type R
1988–1991	ZC	electrical	1988–1991 Civic Si
1992–1995	D15B	electrical	1992–1995 Civic EX and Si
1996–2000	D15B	electrical	1996–2000 Civic EX
1990–1993	F20B	electrical	1992–1995 Prelude Si
1990–1993	F20A	electrical	1992–1995 Prelude Si
1992–1995	F22B	electrical	1992–1995 Prelude Si
1992–1995	H22A	electrical	1993–1995 Prelude Si–VTEC
1996–2001	H22A	electrical	1996–2001 Prelude SH
1988–2000	B16A	mechanical	1994–1997 Del Sol DOHC VTEC
1998–2000	B16B	mechanical	1994–2001 Integra GSR
1996–2000	B20B	mechanical	1996–2000 CRV
1985–1987	ZC	mechanical	1986–1989 Integra
1992–1996	B18B	mechanical	1990–2001 Integra RS and LS
1994–2001	B18C	mechanical	1994–2001 Integra GSR
1996–2000	B18C/R	mechanical	1997–2001 Integra Type R
1988–1991	ZC	mechanical	1986–1989 Integra / 1988–1991 Prelude
1992–2000	D15B	mechanical	1992–2000 Civic EX and Si
1990–1993	F20B	mechanical	1992–1996 Prelude Si
1992–2001	F22B	mechanical	1992–1996 Prelude Si
1992–2001	H22A	mechanical	1993–1996 Prelude Si–VTEC

CHASSIS AND ENGINE CODES

In the process of finding parts and information for your engine swap, you may hear the cars referred to by their chassis code. Here's a quick reference chart to help you remember which cars correspond with which chassis codes.

Honda and Acura Chassis Codes

Vehicle Name	Chassis Code	Vehicle Name	Chassis Code
		1992-1996 Prelude	BB
1984-1987 Civic/CRX	AH/AD		
1988-1991 Civic/ CRX	ED/EE	1986-1989 Integra	DA
1992-1995 Civic/Del Sol	EG/EJ/EH	1990-1993 Integra	DA/DB
1996-2000 Civic	EJ/EM	1994-2001 Integra	DC
2001-2003 Civic	ES/EM/EP	1990-1993 Accord	CB
1988-1991 Prelude	BA	1994-1997 Accord	CD

Honda and Acura Engine Codes

Engine Code	Possible Donor Vehicles	Market Origin
B16A	1990-1991 Civic SiR	JDM
B16A	1990-1991 Civic SiRII	JDM
B16A	1988-1991 Civic SiR hatchback	JDM
B16A	1988-1991 CRX SiR	JDM
B16A	1990-1993 Integra Xsi	JDM
B16A	1990-1993 Integra Rsi	JDM
B16A	1996-2000 Civic SiR	JDM
B16A	1996-2000 Civic SiRII	JDM
B16A	1992-1997 CRX del sol SiR	JDM
B16A	1988-1991 Civic Vti	EDM
B16A	1988-1991 CRX 1.6i	EDM
B16A	1992-1995 Civic SiR	JDM
B16A	1992-1995 Civic Ferio Si	JDM
B16A	1992-1995 Civic Ferio SiII	JDM
B16A	1992-1995 Civic Ferio SiR	JDM
B16A	1992-1995 Civic SiRII	JDM
B16A	1992-1995 Civic Ferio RTSi 4WD	JDM
B16A2	1999-2000 Civic Si	USDM
B16A3	1994-1997 Del Sol Vti-T	EDM
B16A3	1994-1997 Del Sol VTEC	USDM
B16A4	1996-2000 Civic VTi	EDM
B16B	1998-2000 Civic Type R	JDM
B17A1	1992-1993 Integra GSR	USDM
B18A1	1990-1993 Integra RS, LS, GS	USDM
B18B	1992-1996 Domani Si-G	JDM
B18B	1994-1995 Integra Esi	JDM
B18B	1996-2000 Orthia GX	JDM
B18B	1996-2000 Orthia GX-S	JDM
B18B1	1994-2001 Integra RS, LS	USDM
B18C	1996-1997 Integra SiR-G	JDM
B18C	1994-1995 Integra Si-VTEC	JDM
B18C	1996-2000 Integra Type R	JDM
B18C1	1994-2001 Integra GSR	USDM
B18C3	1996 Integra Type R	EDM
B18C4	1994-1995 Integra GSR	EDM
B18C5	1997-2001 Integra Type R (no '99)	USDM
B18C6	1998-2001 Integra Type R	EDM
B20B	1999-2000 S-MX	JDM
B20B	1996-2000 CRV	JDM
B20B	1996-2002 Orthia 2.0GX	JDM
B20B	1996-1998 CRV	USDM

Engine Code	Possible Donor Vehicles	Market Origin
B20B4	1996-2000 CRV	EDM
B20Z	1999-2000 CRV	USDM
D15B	1992-1995 Civic Vti	JDM
D15B	1992-1995 Civic Ferio Vti	JDM
D15B	1992-1997 CRX del sol Vxi	JDM
D15B	1996-2000 Civic Coupe	JDM
D16A1	1986-'87 Integra	USDM

Engine Code	Possible Donor Vehicles	Market Origin
D16A3	1988-1989 Integra	USDM
D16A8	1990-1991 Integra RS, LS	EDM
D16Y8	1996-2000 Civic EX	USDM
D16Z6	1992-1995 Civic EX, SI, Del Sol SI	USDM
F20A	1990-1993 Accord 2.0 Si	JDM
F20A	1990-1993 Ascot 2.0 Si	JDM
F20B	1998-2002 Accord SiR-T	JDM
F20B	1998-2002 Accord SiR	JDM
F22B	1992-1996 Prelude Si	JDM
F22B	1997-2001 Prelude Si	JDM
H22A	1994-1997 Accord SiR	JDM
H22A	1992-1996 Prelude Si-VTEC	JDM
H22A	1997-2001 Prelude Sir	JDM
H22A	1997-2001 Prelude S Spec	JDM
H22A	1997-2001 Prelude Type R	JDM
H22A1	1993-1996 Prelude Si-VTEC	USDM
H22A2	1992-1996 Prelude 2.2i	EDM
H22A4	1997-2001 Prelude SH	USDM
H22A7	1998-2002 Accord Type R	EDM
H23A	1992-1993 Ascot Innova 2.3 Si-Z	JDM
H23A1	1992-1996 Prelude SI	USDM
H23A2	1992-1996 Accord 2.3i	EDM
K20A	2002-2003 RSX	JDM
K20A2	2002-2003 RSX Type S	USDM
K20A3	2002-2003 RSX and Civic SI	USDM
K20C	2002-2003 RSX Type S/Civic Type R	JDM
ZC	1985-1987 Civic Si	JDM
ZC	1985-1987 CRX SI	JDM
ZC	1988-1991 Civic Si	JDM
ZC	1988-1991 CRX Si	JDM
ZC	1986-'89 Integra Gsi	JDM
ZC	1986-'89 Integra Rsi	JDM

If you're thinking about an engine swap, you may be overwhelmed by the variety of engines available. Use this Engine Codes chart to help find out what vehicles have the engine you want. In order to avoid confusion, the vehicles are listed by their US model year. Keep in mind though, that the JDM model years are usually a year earlier (a USDM '94 Integra is actually a '93 JDM Integra).

SWAP
DIFFICULTY

There are various levels of skill required to swap a given engine into a given chassis. Here's a quick reference chart to let you know what you're up against. For example, let's say you have a 1994 Accord and you want to swap in an OBD II H-series engine with a hydraulic manual transmission. As you read the chart from left to right, you see that this swap indicates numbers 1 and 4. When you look these numbers up in the Reference Numbers chart, you'll find that this swap is a beginner bolt-in (1), and only requires a minor electrical conversion (4).

Reference Numbers	
Beginner bolt-in	1
Intermediate bolt-in	2
Fabrication and welding	3
Minor electrical conversion	4
Major electrical conversion	5
Hydro or cable conversion	6
NA	7

Prelude Swaps

Engine	Transmission Type	ECU Type	'88-'91	'92-'96
B series	cable	OBD 0	7	7
B series	cable	OBD I	7	7
B series	cable	OBD II	7	7
B series	hydro	OBD 0	7	7
B series	hydro	OBD I	7	7
B series	hydro	OBD II	7	7
H/F series	hydro	OBD I	1,5	1,4
H/F series	hydro	OBD II	1,5	1,4
ZC (older)	cable	OBD 0	7	7
ZC (newer)	cable	OBD 0	7	7
K series	hydro	OBD II	7	7
SOHC VTEC D series	cable	OBD I	7	7
SOHC VTEC D series	cable	OBD II	7	7
SOHC VTEC D series	hydro	OBD I	7	7
SOHC VTEC D series	hydro	OBD II	7	7

Accord Swaps

Engine	Transmission Type	ECU Type	'90-'93	'94-'97
B series	cable	OBD 0	7	7
B series	cable	OBD I	7	7
B series	cable	OBD II	7	7
B series	hydro	OBD 0	7	7
B series	hydro	OBD I	7	7
B series	hydro	OBD II	7	7
H/F series	hydro	OBD I	1,4	1,4
H/F series	hydro	OBD II	1,4	1,4
ZC (older)	cable	OBD 0	7	7
ZC (newer)	cable	OBD 0	7	7
K series	hydro	OBD II	7	7
SOHC VTEC D series	cable	OBD I	7	7
SOHC VTEC D series	cable	OBD II	7	7
SOHC VTEC D series	hydro	OBD I	7	7
SOHC VTEC D series	hydro	OBD II	7	7

Civic Swaps

Engine	Transmission Type	ECU Type	1984-'87	1988-'91	1992-'95	1996-'00	2001-'03
B series	cable	OBD 0	2,5	2,4	1,5,6	1,5,6	7
B series	cable	OBD I	2,5	2,5	1,4,6	1,5,6	7
B series	cable	OBD II	2,5	2,5	1,5,6	1,4,6	7
B series	hydro	OBD 0	2,5,6	2,4,6	1,5	1,5	7
B series	hydro	OBD I	2,5,6	2,5,6	1,4	1,5	7
B series	hydro	OBD II	2,5,6	2,5,6	1,5	1,4	7
H/F series	hydro	OBD I	3,5,6	3,5,6	2,5	2,5	7
H/F series	hydro	OBD II	3,5,6	3,5,6	2,5	2,5	7
ZC (older)	cable	OBD 0	1,4	7	7	7	7
ZC (newer)	cable	OBD 0	7	1,4	7	7	7
K series	hydro	OBD II	3,5,6	3,5,6	3,5	3,5	2,5
SOHC VTEC D series	cable	OBD I	3,5	1,5	1,4,6	1,5,6	7
SOHC VTEC D series	cable	OBD II	3,5	1,5	1,5,6	1,4,6	7
SOHC VTEC D series	hydro	OBD I	3,5,6	1,5,6	1,4	1,5	7
SOHC VTEC D series	hydro	OBD II	3,5,6	1,5,6	1,5	1,4	7

Integra Swaps

Engine	Transmission Type	ECU Type	1986-'89	1990-'93	1994-'01
B series	cable	OBD 0	2,5	1,4	1,5,6
B series	cable	OBD I	2,5	1,4	1,4,6
B series	cable	OBD II	2,5	1,5	1,4,6
B series	hydro	OBD 0	2,5,6	1,4,6	1,5
B series	hydro	OBD I	2,5,6	1,4,6	1,4
B series	hydro	OBD II	2,5,6	1,5,6	1,4
H/F series	hydro	OBD I	3,5,6	2,5,6	2,5
H/F series	hydro	OBD II	3,5,6	2,5,6	2,5
ZC (older)	cable	OBD 0	1,4	7	7
ZC (newer)	cable	OBD 0	7	7	7
K series	hydro	OBD II	3,5,6	3,5,6	3,5
SOHC VTEC D series	cable	OBD I	3,5	7	7
SOHC VTEC D series	cable	OBD II	3,5	7	7
SOHC VTEC D series	hydro	OBD I	3,5,6	7	7
SOHC VTEC D series	hydro	OBD II	3,5,6	7	7

Source Guide

HCP Engineering
5179 Brooks Street, Suite H
Montclair, CA 91763
909-399-3400
www.hcpengineering.com

Hasport Performance
4046 East Winslow Avenue
Phoenix, AZ 85040
602-470-0065
www.hasport.com

Place Racing
1016 West Gladstone Street
Azusa, CA 91702
626-334-3345
www.placeracing.com

Go Motors
1271 North Sunshine Way
Anaheim, CA 92806
1-888-4GO-MOTOR
www.japaneseusedengines.com

Progress Suspension
1390 North Hundley Street
Anaheim, CA 92806-1301
800-905-6687
www.progressauto.com

Tustin Acura Parts Department
9 Auto Center Drive
Tustin, CA 92782
714-699-9900
www.tustinacura.com

MORE GREAT TITLES AVAILABLE FROM CARTECH®

CHEVROLET

How To Rebuild the Small-Block Chevrolet* *(SA26)*
Chevrolet Small-Block Parts Interchange Manual *(SA55)*
How To Build Max Perf Chevy Small-Blocks on a Budget *(SA57)*
How To Build High-Perf Chevy LS1/LS6 Engines *(SA86)*
How To Build Big-Inch Chevy Small-Blocks *(SA87)*
How to Build High-Performance Chevy Small-Block Cams/Valvetrains *SA105*
Rebuilding the Small-Block Chevy: Step-by-Step Videobook *(SA116)*
High-Performance Chevy Small-Block Cylinder Heads *(SA125P)*
High Performance C5 Corvette Builder's Guide *(SA127)*
How to Rebuild the Big-Block Chevrolet* *(SA142P)*
How to Build Max-Performance Chevy Big Block on a Budget *(SA198)*
How to Restore Your Camaro 1967–1969 *(SA178)*
How to Build Killer Big-Block Chevy Engines *(SA190)*
How to Build Max-Performance Chevy LT1/LT4 Engines *(SA206)*
Small-Block Chevy Performance: 1955-1996 *(SA110P)*
How to Build Small-Block Chevy Circle-Track Racing Engines *(SA121P)*
High-Performance C5 Corvette Builder's Guide *(SA127P)*
Chevrolet Big Block Parts Interchange Manual *(SA31P)*
Chevy TPI Fuel Injection Swapper's Guide *(SA53P)*

FORD

High-Performance Ford Engine Parts Interchange *(SA56)*
How To Build Max Performance Ford V-8s on a Budget *(SA69)*
How To Build Max Perf 4.6 Liter Ford Engines *(SA82)*
How To Build Big-Inch Ford Small-Blocks *(SA85)*
How to Rebuild the Small-Block Ford* *(SA102)*
How to Rebuild Big-Block Ford Engines* *(SA162)*
Full-Size Fords 1955–1970 *(SA176)*
How to Build Max-Performance Ford FE Engines *(SA183)*
How to Restore Your Mustang 1964 1/2–1973 *(SA165)*
How to Build Ford RestoMod Street Machines *(SA101P)*
Building 4.6/5.4L Ford Horsepower on the Dyno *(SA115P)*
How to Rebuild 4.6/5.4-Liter Ford Engines* *(SA155P)*
Building High-Performance Fox-Body Mustangs on a Budget *(SA75P)*
How to Build Supercharged & Turbocharged Small-Block Fords *(SA95P)*

GENERAL MOTORS

GM Automatic Overdrive Transmission Builder's and Swapper's Guide *(SA140)*
How to Rebuild GM LS-Series Engines* *(SA147)*
How to Swap GM LS-Series Engines Into Almost Anything *(SA156)*
How to Supercharge & Turbocharge GM LS-Series Engines *(SA180)*
How to Build Big-Inch GM LS-Series Engines *(SA203)*
How to Rebuild & Modify GM Turbo 400 Transmissions* *(SA186)*
How to Build GM Pro-Touring Street Machines *(SA81P)*

MOPAR

How to Rebuild the Big-Block Mopar* *(SA197)*
How to Rebuild the Small-Block Mopar* *(SA143P)*
How to Build Max-Performance Hemi Engines *(SA164)*
How To Build Max-Performance Mopar Big Blocks *(SA171)*
Mopar B-Body Performance Upgrades 1962-1979 *(SA191)*
How to Build Big-Inch Mopar Small-Blocks *(SA104P)*
High-Performance New Hemi Builder's Guide 2003-Present *(SA132P)*

OLDSMOBILE/ PONTIAC/ BUICK

How to Build Max-Performance Oldsmobile V-8s *(SA172)*
How To Build Max-Perf Pontiac V-8s *SA78*
How to Rebuild Pontiac V-8s* *(SA200)*
How to Build Max-Performance Buick Engines *(SA146P)*

SPORT COMPACTS

Honda Engine Swaps *(SA93)*
Building Honda K-Series Engine Performance *(SA134)*
High-Performance Subaru Builder's Guide *(SA141)*
How to Build Max-Performance Mitsubishi 4G63t Engines *(SA148)*
How to Rebuild Honda B-Series Engines* *(SA154)*
The New Mini Performance Handbook *(SA182P)*
High Performance Dodge Neon Builder's Handbook *(SA100P)*
High-Performance Honda Builder's Handbook Volume 1 *(SA49P)*

Workbench® Series books feature step-by-step instruction with hundreds of color photos for stock rebuilds and automotive repair.

ENGINE

Engine Blueprinting *(SA21)*
Automotive Diagnostic Systems: Understanding OBD-I & OBD II *(SA174)*

INDUCTION & IGNITION

Super Tuning & Modifying Holley Carburetors *(SA08)*
Street Supercharging, A Complete Guide to *(SA17)*
How To Build High-Performance Ignition Systems *(SA79)*
How to Build and Modify Rochester Quadrajet Carburetors *(SA113)*
Turbo: Real World High-Performance Turbocharger Systems *(SA123)*
How to Rebuild & Modify Carter/Edelbrock Carbs *(SA130)*
Engine Management: Advanced Tuning *(SA135)*
Designing & Tuning High-Performance Fuel Injection Systems *(SA161)*
Demon Carburetion *(SA68P)*

DRIVING

How to Drift: The Art of Oversteer *(SA118P)*
How to Drag Race *(SA136)*
How to Autocross *(SA158P)*
How to Hook and Launch *(SA195)*

HIGH-PERFORMANCE & RESTORATION HOW-TO

How To Install and Tune Nitrous Oxide Systems *(SA194)*
Custom Painting *(SA10)*
David Vizard's How to Build Horsepower *(SA24)*
How to Rebuild & Modify High-Performance Manual Transmissions* *(SA103)*
High-Performance Jeep Cherokee XJ Builder's Guide 1984–2001 *(SA109)*
How to Paint Your Car on a Budget *(SA117)*
High Performance Brake Systems *(SA126P)*
High Performance Diesel Builder's Guide *(SA129)*
4x4 Suspension Handbook *(SA137)*
How to Rebuild Any Automotive Engine* *(SA151)*
Automotive Welding: A Practical Guide* *(SA159)*
Automotive Wiring and Electrical Systems* *(SA160)*
Design & Install In-Car Entertainment Systems *(SA163)*
Automotive Bodywork & Rust Repair* *(SA166)*
High-Performance Differentials, Axles, & Drivelines *(SA170)*
How to Make Your Muscle Car Handle *(SA175)*
Rebuilding Any Automotive Engine: Step-by-Step Videobook *(SA179)*
Builder's Guide to Hot Rod Chassis & Suspension *(SA185)*
How To Rebuild & Modify GM Turbo 400 Transmissions* *(SA186)*
How to Build Altered Wheelbase Cars *(SA189)*
How to Build Period Correct Hot Rods *(SA192)*
Automotive Sheet Metal Forming & Fabrication *(SA196)*
Performance Automotive Engine Math *(SA204)*
How to Design, Build & Equip Your Automotive Workshop on a Budget *(SA207)*
Automotive Electrical Performance Projects *(SA209)*
How to Port Cylinder Heads *(SA215)*
Muscle Car Interior Restoration Guide *(SA167)*
High Performance Jeep Wrangler TJ Builder's Guide: 1997-2006 *(SA120P)*
Dyno Testing & Tuning *(SA138P)*
How to Rebuild Any Automotive Engine *(SA151P)*
Muscle Car Interior Restoration Guide *(SA167P)*
How to Build Horsepower - Volume 2 *(SA52P)*
Bolt-Together Street Rods *(SA72P)*

HISTORIES & PERSONALITIES

Fuelies: Fuel Injected Corvettes 1957–1965 *(CT452)*
Yenko *(CT485)*
Lost Hot Rods *(CT487)*
Grumpy's Toys *(CT489)*
Rusted Muscle — A collection of junkyard muscle cars. *(CT492)*
America's Coolest Station Wagons *(CT493)*
Super Stock — A paperback version of a classic best seller. *(CT495)*
Rusty Pickups: American Workhorses Put to Pasture *(CT496)*
Jerry Heasley's Rare Finds — Great collection of Heasley's best finds. *(CT497)*
Street Sleepers: The Art of the Deceptively Fast Car *(CT498)*
Ed 'Big Daddy' Roth — Paperback reprint of a classic best seller. *(CT500)*
Rat Rods: Rodding's Imperfect Stepchildren *(CT486)*
East vs. West Showdown: Rods, Customs Rails *(CT501)*
Car Spy: Secret Cars Exposed by the Industry's Most Notorious Photographer *(CT502)*

CarTech®, Inc. 39966 Grand Ave., North Branch, MN 55056. Ph: 800-551-4754 or 651-277-1200 • Fax: 651-277-1203
Brooklands Books Ltd., PO Box 146 Cobham, Surrey KT11 1LG, England. Ph: 01932 865051 • Fax 01932 868803
Brooklands Books Aus., 3/37-39 Green Street, Banksmeadow, NSW 2019, Australia. Ph: 2 9695 7055 • Fax 2 9695 7355

Visit us online at
www.cartechbooks.com for more info!

Additional books that may interest you...

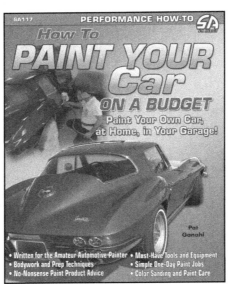

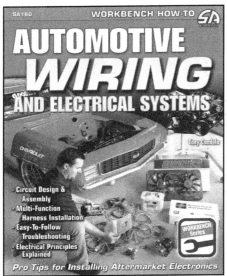

CPSIA information can be obtained
at www.ICGtesting.com
Printed in the USA
BVHW010140200619
551448BV00008B/215/P